Developmental Biology: A Very Short Introduction

VERY SHORT INTRODUCTIONS are for anyone wanting a stimulating and accessible way into a new subject. They are written by experts, and have been translated into more than 45 different languages.

The series began in 1995, and now covers a wide variety of topics in every discipline. The VSI library now contains over 500 volumes—a Very Short Introduction to everything from Psychology and Philosophy of Science to American History and Relativity—and continues to grow in every subject area.

Titles in the series include the following:

AFRICAN HISTORY John Parker and Richard Rathbone
AGEING Nancy A. Pachana
AGNOSTICISM Robin Le Poidevin
AGRICULTURE Paul Brassley and Richard Soffe
ALEXANDER THE GREAT Hugh Bowden
ALGEBRA Peter M. Higgins
AMERICAN HISTORY Paul S. Boyer
AMERICAN IMMIGRATION David A. Gerber
AMERICAN LEGAL HISTORY G. Edward White
AMERICAN POLITICAL HISTORY Donald Critchlow
AMERICAN POLITICAL PARTIES AND ELECTIONS L. Sandy Maisel
AMERICAN POLITICS Richard M. Valelly
THE AMERICAN PRESIDENCY Charles O. Jones
AMERICAN SLAVERY Heather Andrea Williams
THE AMERICAN WEST Stephen Aron
AMERICAN WOMEN'S HISTORY Susan Ware
ANAESTHESIA Aidan O'Donnell
ANARCHISM Colin Ward
ANCIENT EGYPT Ian Shaw
ANCIENT GREECE Paul Cartledge
THE ANCIENT NEAR EAST Amanda H. Podany
ANCIENT PHILOSOPHY Julia Annas
ANCIENT WARFARE Harry Sidebottom
ANGLICANISM Mark Chapman
THE ANGLO-SAXON AGE John Blair
ANIMAL BEHAVIOUR Tristram D. Wyatt
ANIMAL RIGHTS David DeGrazia
ANXIETY Daniel Freeman and Jason Freeman
ARCHAEOLOGY Paul Bahn
ARISTOTLE Jonathan Barnes
ART HISTORY Dana Arnold
ART THEORY Cynthia Freeland
ASTROPHYSICS James Binney
ATHEISM Julian Baggini
THE ATMOSPHERE Paul I. Palmer
AUGUSTINE Henry Chadwick
THE AZTECS Davíd Carrasco
BABYLONIA Trevor Bryce
BACTERIA Sebastian G. B. Amyes
BANKING John Goddard and John O. S. Wilson
BARTHES Jonathan Culler
BEAUTY Roger Scruton
THE BIBLE John Riches
BLACK HOLES Katherine Blundell
BLOOD Chris Cooper
THE BODY Chris Shilling
THE BOOK OF MORMON Terryl Givens
BORDERS Alexander C. Diener and Joshua Hagen
THE BRAIN Michael O'Shea
THE BRICS Andrew F. Cooper
BRITISH POLITICS Anthony Wright

BUDDHA Michael Carrithers
BUDDHISM Damien Keown
BUDDHIST ETHICS Damien Keown
BYZANTIUM Peter Sarris
CANCER Nicholas James
CAPITALISM James Fulcher
CATHOLICISM Gerald O'Collins
CAUSATION Stephen Mumford and Rani Lill Anjum
THE CELL Terence Allen and Graham Cowling
THE CELTS Barry Cunliffe
CHEMISTRY Peter Atkins
CHILD PSYCHOLOGY Usha Goswami
CHINESE LITERATURE Sabina Knight
CHOICE THEORY Michael Allingham
CHRISTIAN ART Beth Williamson
CHRISTIAN ETHICS D. Stephen Long
CHRISTIANITY Linda Woodhead
CIRCADIAN RHYTHMS Russell Foster and Leon Kreitzman
CITIZENSHIP Richard Bellamy
CLASSICAL MYTHOLOGY Helen Morales
CLASSICS Mary Beard and John Henderson
CLIMATE Mark Maslin
CLIMATE CHANGE Mark Maslin
COGNITIVE NEUROSCIENCE Richard Passingham
THE COLD WAR Robert McMahon
COLONIAL AMERICA Alan Taylor
COLONIAL LATIN AMERICAN LITERATURE Rolena Adorno
COMBINATORICS Robin Wilson
COMMUNISM Leslie Holmes
COMPLEXITY John H. Holland
THE COMPUTER Darrel Ince
COMPUTER SCIENCE Subrata Dasgupta
CONFUCIANISM Daniel K. Gardner
CONSCIOUSNESS Susan Blackmore
CONTEMPORARY ART Julian Stallabrass
CONTEMPORARY FICTION Robert Eaglestone
CONTINENTAL PHILOSOPHY Simon Critchley
CORAL REEFS Charles Sheppard
CORPORATE SOCIAL RESPONSIBILITY Jeremy Moon
COSMOLOGY Peter Coles
CRIMINAL JUSTICE Julian V. Roberts
CRITICAL THEORY Stephen Eric Bronner
THE CRUSADES Christopher Tyerman
CRYSTALLOGRAPHY A. M. Glazer
DADA AND SURREALISM David Hopkins
DANTE Peter Hainsworth and David Robey
DARWIN Jonathan Howard
THE DEAD SEA SCROLLS Timothy Lim
DECOLONIZATION Dane Kennedy
DEMOCRACY Bernard Crick
DEPRESSION Jan Scott and Mary Jane Tacchi
DERRIDA Simon Glendinning
DESIGN John Heskett
DEVELOPMENTAL BIOLOGY Lewis Wolpert
DIASPORA Kevin Kenny
DINOSAURS David Norman
DREAMING J. Allan Hobson
DRUGS Les Iversen
DRUIDS Barry Cunliffe
THE EARTH Martin Redfern
EARTH SYSTEM SCIENCE Tim Lenton
ECONOMICS Partha Dasgupta
EGYPTIAN MYTH Geraldine Pinch
EIGHTEENTH-CENTURY BRITAIN Paul Langford
THE ELEMENTS Philip Ball
EMOTION Dylan Evans
EMPIRE Stephen Howe
ENGLISH LITERATURE Jonathan Bate
THE ENLIGHTENMENT John Robertson
ENVIRONMENTAL ECONOMICS Stephen Smith
ENVIRONMENTAL POLITICS Andrew Dobson
EPICUREANISM Catherine Wilson
EPIDEMIOLOGY Rodolfo Saracci
ETHICS Simon Blackburn
THE ETRUSCANS Christopher Smith
EUGENICS Philippa Levine
THE EUROPEAN UNION John Pinder and Simon Usherwood
EVOLUTION Brian and Deborah Charlesworth

EXISTENTIALISM Thomas Flynn
THE EYE Michael Land
FAMILY LAW Jonathan Herring
FASCISM Kevin Passmore
FEMINISM Margaret Walters
FILM Michael Wood
FILM MUSIC Kathryn Kalinak
THE FIRST WORLD WAR Michael Howard
FOOD John Krebs
FORENSIC PSYCHOLOGY David Canter
FORENSIC SCIENCE Jim Fraser
FORESTS Jaboury Ghazoul
FOSSILS Keith Thomson
FOUCAULT Gary Gutting
FREE SPEECH Nigel Warburton
FREE WILL Thomas Pink
FREUD Anthony Storr
FUNDAMENTALISM Malise Ruthven
FUNGI Nicholas P. Money
THE FUTURE Jennifer M. Gidley
GALAXIES John Gribbin
GALILEO Stillman Drake
GAME THEORY Ken Binmore
GANDHI Bhikhu Parekh
GEOGRAPHY John Matthews and David Herbert
GEOPOLITICS Klaus Dodds
GERMAN LITERATURE Nicholas Boyle
GERMAN PHILOSOPHY Andrew Bowie
GLOBAL CATASTROPHES Bill McGuire
GLOBAL ECONOMIC HISTORY Robert C. Allen
GLOBALIZATION Manfred Steger
GOD John Bowker
GRAVITY Timothy Clifton
THE GREAT DEPRESSION AND THE NEW DEAL Eric Rauchway
HABERMAS James Gordon Finlayson
THE HEBREW BIBLE AS LITERATURE Tod Linafelt
HEGEL Peter Singer
HERODOTUS Jennifer T. Roberts
HIEROGLYPHS Penelope Wilson
HINDUISM Kim Knott
HISTORY John H. Arnold
THE HISTORY OF ASTRONOMY Michael Hoskin
THE HISTORY OF CHEMISTRY William H. Brock
THE HISTORY OF LIFE Michael Benton
THE HISTORY OF MATHEMATICS Jacqueline Stedall
THE HISTORY OF MEDICINE William Bynum
THE HISTORY OF TIME Leofranc Holford-Strevens
HIV AND AIDS Alan Whiteside
HOLLYWOOD Peter Decherney
HUMAN ANATOMY Leslie Klenerman
HUMAN EVOLUTION Bernard Wood
HUMAN RIGHTS Andrew Clapham
THE ICE AGE Jamie Woodward
IDEOLOGY Michael Freeden
INDIAN PHILOSOPHY Sue Hamilton
THE INDUSTRIAL REVOLUTION Robert C. Allen
INFECTIOUS DISEASE Marta L. Wayne and Benjamin M. Bolker
INFINITY Ian Stewart
INFORMATION Luciano Floridi
INNOVATION Mark Dodgson and David Gann
INTELLIGENCE Ian J. Deary
INTERNATIONAL MIGRATION Khalid Koser
INTERNATIONAL RELATIONS Paul Wilkinson
IRAN Ali M. Ansari
ISLAM Malise Ruthven
ISLAMIC HISTORY Adam Silverstein
ISOTOPES Rob Ellam
ITALIAN LITERATURE Peter Hainsworth and David Robey
JESUS Richard Bauckham
JOURNALISM Ian Hargreaves
JUDAISM Norman Solomon
JUNG Anthony Stevens
KABBALAH Joseph Dan
KANT Roger Scruton
KNOWLEDGE Jennifer Nagel
THE KORAN Michael Cook
LATE ANTIQUITY Gillian Clark
LAW Raymond Wacks
THE LAWS OF THERMODYNAMICS Peter Atkins
LEADERSHIP Keith Grint
LEARNING Mark Haselgrove

LEIBNIZ Maria Rosa Antognazza
LIBERALISM Michael Freeden
LIGHT Ian Walmsley
LINGUISTICS Peter Matthews
LITERARY THEORY Jonathan Culler
LOCKE John Dunn
LOGIC Graham Priest
MACHIAVELLI Quentin Skinner
MAGIC Owen Davies
MAGNA CARTA Nicholas Vincent
MAGNETISM Stephen Blundell
MARINE BIOLOGY Philip V. Mladenov
MARTIN LUTHER Scott H. Hendrix
MARTYRDOM Jolyon Mitchell
MARX Peter Singer
MATERIALS Christopher Hall
MATHEMATICS Timothy Gowers
THE MEANING OF LIFE Terry Eagleton
MEASUREMENT David Hand
MEDICAL ETHICS Tony Hope
MEDIEVAL BRITAIN John Gillingham and Ralph A. Griffiths
MEDIEVAL LITERATURE Elaine Treharne
MEDIEVAL PHILOSOPHY John Marenbon
MEMORY Jonathan K. Foster
METAPHYSICS Stephen Mumford
MICROBIOLOGY Nicholas P. Money
MICROECONOMICS Avinash Dixit
MICROSCOPY Terence Allen
THE MIDDLE AGES Miri Rubin
MILITARY JUSTICE Eugene R. Fidell
MINERALS David Vaughan
MODERN ART David Cottington
MODERN CHINA Rana Mitter
MODERN FRANCE Vanessa R. Schwartz
MODERN IRELAND Senia Pašeta
MODERN ITALY Anna Cento Bull
MODERN JAPAN Christopher Goto-Jones
MODERNISM Christopher Butler
MOLECULAR BIOLOGY Aysha Divan and Janice A. Royds
MOLECULES Philip Ball
MOONS David A. Rothery
MOUNTAINS Martin F. Price
MUHAMMAD Jonathan A. C. Brown
MUSIC Nicholas Cook
MYTH Robert A. Segal
THE NAPOLEONIC WARS Mike Rapport
NELSON MANDELA Elleke Boehmer
NEOLIBERALISM Manfred Steger and Ravi Roy
NEWTON Robert Iliffe
NIETZSCHE Michael Tanner
NINETEENTH-CENTURY BRITAIN Christopher Harvie and H. C. G. Matthew
NORTH AMERICAN INDIANS Theda Perdue and Michael D. Green
NORTHERN IRELAND Marc Mulholland
NOTHING Frank Close
NUCLEAR PHYSICS Frank Close
NUMBERS Peter M. Higgins
NUTRITION David A. Bender
THE OLD TESTAMENT Michael D. Coogan
ORGANIC CHEMISTRY Graham Patrick
THE PALESTINIAN-ISRAELI CONFLICT Martin Bunton
PANDEMICS Christian W. McMillen
PARTICLE PHYSICS Frank Close
THE PERIODIC TABLE Eric R. Scerri
PHILOSOPHY Edward Craig
PHILOSOPHY IN THE ISLAMIC WORLD Peter Adamson
PHILOSOPHY OF LAW Raymond Wacks
PHILOSOPHY OF SCIENCE Samir Okasha
PHOTOGRAPHY Steve Edwards
PHYSICAL CHEMISTRY Peter Atkins
PILGRIMAGE Ian Reader
PLAGUE Paul Slack
PLANETS David A. Rothery
PLANTS Timothy Walker
PLATE TECTONICS Peter Molnar
PLATO Julia Annas
POLITICAL PHILOSOPHY David Miller
POLITICS Kenneth Minogue
POPULISM Cas Mudde and Cristóbal Rovira Kaltwasser
POSTCOLONIALISM Robert Young

POSTMODERNISM Christopher Butler
POSTSTRUCTURALISM
Catherine Belsey
PREHISTORY Chris Gosden
PRESOCRATIC PHILOSOPHY
Catherine Osborne
PRIVACY Raymond Wacks
PSYCHIATRY Tom Burns
PSYCHOLOGY Gillian Butler and
Freda McManus
PSYCHOTHERAPY Tom Burns and
Eva Burns-Lundgren
PUBLIC ADMINISTRATION
Stella Z. Theodoulou and Ravi K. Roy
PUBLIC HEALTH Virginia Berridge
QUANTUM THEORY
John Polkinghorne
RACISM Ali Rattansi
REALITY Jan Westerhoff
THE REFORMATION Peter Marshall
RELATIVITY Russell Stannard
RELIGION IN AMERICA Timothy Beal
THE RENAISSANCE Jerry Brotton
RENAISSANCE ART
Geraldine A. Johnson
REVOLUTIONS Jack A. Goldstone
RHETORIC Richard Toye
RISK Baruch Fischhoff and John Kadvany
RITUAL Barry Stephenson
RIVERS Nick Middleton
ROBOTICS Alan Winfield
ROMAN BRITAIN Peter Salway
THE ROMAN EMPIRE Christopher Kelly
THE ROMAN REPUBLIC
David M. Gwynn
RUSSIAN HISTORY Geoffrey Hosking
RUSSIAN LITERATURE Catriona Kelly
THE RUSSIAN REVOLUTION
S. A. Smith
SAVANNAS Peter A. Furley
SCHIZOPHRENIA Chris Frith and
Eve Johnstone
SCIENCE AND RELIGION
Thomas Dixon
THE SCIENTIFIC REVOLUTION
Lawrence M. Principe
SCOTLAND Rab Houston
SEXUALITY Véronique Mottier
SHAKESPEARE'S COMEDIES Bart van Es
SIKHISM Eleanor Nesbitt
SLEEP Steven W. Lockley and
Russell G. Foster
SOCIAL AND CULTURAL
ANTHROPOLOGY
John Monaghan and Peter Just
SOCIAL PSYCHOLOGY Richard J. Crisp
SOCIAL WORK Sally Holland and
Jonathan Scourfield
SOCIALISM Michael Newman
SOCIOLOGY Steve Bruce
SOCRATES C. C. W. Taylor
SOUND Mike Goldsmith
THE SOVIET UNION Stephen Lovell
THE SPANISH CIVIL WAR
Helen Graham
SPANISH LITERATURE Jo Labanyi
SPORT Mike Cronin
STARS Andrew King
STATISTICS David J. Hand
STUART BRITAIN John Morrill
SYMMETRY Ian Stewart
TAXATION Stephen Smith
TELESCOPES Geoff Cottrell
TERRORISM Charles Townshend
THEOLOGY David F. Ford
TIBETAN BUDDHISM
Matthew T. Kapstein
THE TROJAN WAR Eric H. Cline
THE TUDORS John Guy
TWENTIETH-CENTURY BRITAIN
Kenneth O. Morgan
THE UNITED NATIONS
Jussi M. Hanhimäki
THE U.S. CONGRESS Donald A. Ritchie
THE U.S. SUPREME COURT
Linda Greenhouse
THE VIKINGS Julian Richards
VIRUSES Dorothy H. Crawford
VOLTAIRE Nicholas Cronk
WAR AND TECHNOLOGY
Alex Roland
WATER John Finney
WILLIAM SHAKESPEARE
Stanley Wells
WITCHCRAFT Malcolm Gaskill
THE WORLD TRADE
ORGANIZATION Amrita Narlikar
WORLD WAR II Gerhard L. Weinberg

Lewis Wolpert

DEVELOPMENTAL BIOLOGY

A Very Short Introduction

OXFORD
UNIVERSITY PRESS

Great Clarendon Street, Oxford OX2 6DP

Oxford University Press is a department of the University of Oxford.
It furthers the University's objective of excellence in research, scholarship,
and education by publishing worldwide in

Oxford New York

Auckland Cape Town Dar es Salaam Hong Kong Karachi
Kuala Lumpur Madrid Melbourne Mexico City Nairobi
New Delhi Shanghai Taipei Toronto

With offices in

Argentina Austria Brazil Chile Czech Republic France Greece
Guatemala Hungary Italy Japan Poland Portugal Singapore
South Korea Switzerland Thailand Turkey Ukraine Vietnam

Oxford is a registered trade mark of Oxford University Press
in the UK and in certain other countries

Published in the United States
by Oxford University Press Inc., New York

First published 2011

British Library Cataloguing in Publication Data
Data available

Library of Congress Cataloging in Publication Data
Data available

Typeset by SPI Publisher Services, Pondicherry, India
Printed and bound by
CPI Group (UK) Ltd, Croydon, CR0 4YY

ISBN: 978-0-19-960119-6

Contents

List of illustrations

All illustrations from Wolpert, L. and Tickle, C. 2010 *Principles of Development* 4th edn Oxford University Press, unless otherwise stated.

Introduction

That we develop from a single cell, the fertilized egg, just one tenth of a millimetre in diameter— smaller than a full stop—is amazing. That egg has all the information to develop into a human being. Though many of the mechanisms are understood, there are still many uncertainties.

The embryo, the name now given to the structure to which the dividing egg gives rise, hid its achievement for a very long time. A scientific approach to explaining development of the embryo started with Hippocrates in Greece in the 5th century BC. Using the ideas current at the time, he tried to explain development in terms of heat, wetness, and solidification. About a century later, the Greek philosopher Aristotle formulated a question that was to dominate much thinking about development until the end of the 19th century. He considered two possibilities: one was that everything in the embryo was preformed from the very beginning and simply got bigger during development; the other, which he favoured, was that new structures arose progressively, a process he termed epigenesis and that he likened metaphorically to the 'knitting of a net'. His ideas remained dominant well into the 17th century when the contrary view, namely that the embryo was preformed from the beginning, was championed. Many could not believe that physical or chemical forces could mould a living entity like we humans from the embryo. Along with the belief in the divine creation of the world and all living things, was the belief that all embryos had existed from the beginning of the world. But the problem could not be resolved until one of the great advances in biology had taken place at the end of the 19th century—the recognition that living things, including embryos, were composed of cells and that the embryo developed from a single cell, the egg. All the cells in the adult come from that fertilized egg dividing many times. Another important advance was the proposal by the German biologist August Weismann that the offspring does not inherit its characteristics from the body of the parent, but only from the germ cells—eggs and sperm. Then later came the discovery of DNA and genes and how they code for proteins which, in turn, determine how cells behave.

A big problem was raised by the experiment of Hans Driesch more than 100 years ago when he separated the two cells of the sea urchin embryo after the first division and each developed into a normal but smaller larva (Figure 1). This early embryo thus had flag-like properties—its pattern was the same over a range of sizes. It was also the first clear demonstration of the developmental process known as **regulation**—the ability of an embryo to restore normal development even if some portions are removed or rearranged very early in development—and showed that the fate of cells is not fixed at an early stage. This ability to develop normally even if the early embryo is smaller also applies to human identical twins when the early embryo splits into two.

The fact that embryos can regulate their development implies that cells must interact with each other, but the central importance of cell–cell interactions in embryonic development was not really established until the discovery of the phenomenon of **induction**. This is where one group of cells directs the development of a neighbouring cell or tissue. The importance of induction and other cell–cell interactions in development was proved dramatically in 1924 when Hans Spemann and his assistant Hilde Mangold carried out a famous transplantation experiment in amphibian

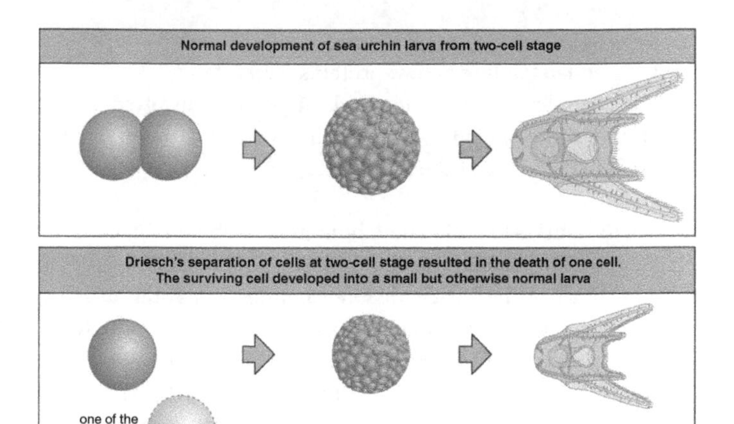

1. Driesch's experiment on sea urchin embryos, which first demonstrated the phenomenon of regulation. After separation of cells at the two-cell stage, one cell usually died and the remaining cell developed into a small, but whole, normal larva

embryos. They showed that a partial second embryo could be induced by grafting one small region of an early newt embryo onto another embryo at the same stage. That region is now known as the Spemann organizer.

The development of multicellular organisms from the fertilized egg is a brilliant triumph of evolution. The human fertilized egg divides to give rise to many millions of cells, which form structures as complex and varied as eyes, arms, heart, and brain. This amazing achievement raises a multitude of questions. How do the cells arising from division of the fertilized egg become different from each other? How do they become organized into structures such as limbs and brains? What controls the behaviour of individual cells so that such highly organized patterns emerge? How are the organizing principles of development embedded within the egg and, in particular, within the genes? Much of the excitement in developmental biology today comes from our growing understanding of how proteins direct these developmental processes—thousands of genes are involved in controlling development by determining what proteins are made in the right place and at the right time.

One of the aims is to understand human development in order to understand why it sometimes goes wrong, why a foetus may fail to be born, or a baby be born with abnormalities. Mutations in genes can lead to abnormal development, as can environmental factors, such as drugs and infections. Another area of medical research related to developmental biology is regenerative medicine—finding out how to use cells to repair damaged tissues and organs. The focus of regenerative medicine is currently on **stem cells**, which have many of the properties of embryonic cells, such as the ability to proliferate and to develop into a range of different tissues.

A relatively small number of animals have been chosen for intensive study of development of the embryo because they were amenable to experimental manipulation, or genetic analysis. This is why the frog *Xenopus laevis*, the nematode worm *Caenorhabditis elegans*, the fruit fly *Drosophila melanogaster*, the zebrafish, the chick, and the mouse have such a dominant place in developmental biology. Similarly, work with the thale cress, *Arabidopsis thaliana*, has uncovered many features of plant development. Understanding a developmental process in one organism can help to illuminate similar processes elsewhere. For example, the identification of genes controlling early embryogenesis in the fly has led to the discovery of related genes being used in similar ways in the development of vertebrates including humans. Each species has its advantages and disadvantages as a developmental model. The fly has been wonderful for genetics. Frog and chick embryos are robust for surgical manipulation, and easily accessible to the experimenter at all stages in their development, unlike those of mammals. Chick embryos are very similar to those of mammals in the general course of embryonic development, but are easier to manipulate. Many observations can be carried out simply by making a window in the eggshell, and the embryo can also be cultured outside the egg. Mouse development is hidden from view and can only be followed by isolating embryos at different stages. The mouse has nevertheless become the main model organism for mammalian development and was the first mammal after humans to have its complete genome sequenced and it is widely used for genetic studies. The zebrafish is a more recent addition to the select list of vertebrate model systems; it is easy to breed in large numbers, the embryos are transparent and so cell divisions and tissue movements can be followed visually, and it has great potential for genetic investigations. The nematode worm has the great advantage of having a fixed number of cells, 959, and the development of each can be followed.

An example illustrating some of the main features of vertebrate development is the frog (Figure 2). The unfertilized egg is a large cell because it contains much yolk. Fertilization of the egg by a sperm is followed by fusion of male and female nuclei, after which **cleavage** begins. Cleavages are divisions in which cells do not grow between each division, and so with successive cleavages the cells become smaller. After about 12 division cycles, the embryo, now known as a blastula, consists of many small cells surrounding a fluid-filled cavity above the larger yolky cells. These give rise to the three germ layers that form the embryo—ectoderm, endoderm, and mesoderm—all of which are still on the surface of the embryo. The upper region, ectoderm, will form both the epidermis of the skin and the nervous system; endoderm gives rise to the gut; and mesoderm will give rise to the internal structures like the skeleton. During the next stage—**gastrulation**—there is a dramatic rearrangement of cells. The endoderm and mesoderm move inside through a small region known as the blastopore and the basic body plan of the tadpole is established. The ectoderm remains on the outside.

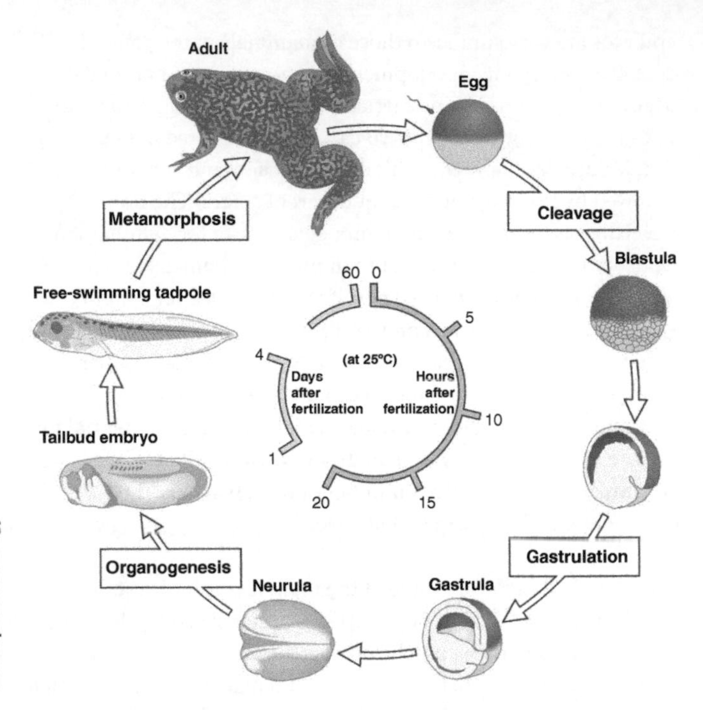

2. Life cycle of the African claw-toed frog *Xenopus laevis*

Internally, the mesoderm gives rise to a rod-like structure, the notochord, which runs from the head to the tail; the nervous system will later develop above the notochord, which then disappears. On either side of the notochord are segmented blocks of mesoderm called somites, which will give rise to the muscles and vertebral column. Shortly after gastrulation, the ectoderm above the notochord folds to form the neural tube, which gives rise to the brain and spinal cord—a process known as **neurulation**. By this time, other organs, such as limbs, eyes, are specified at their future locations, but only develop a little later, during organogenesis. During organogenesis, specialized cells

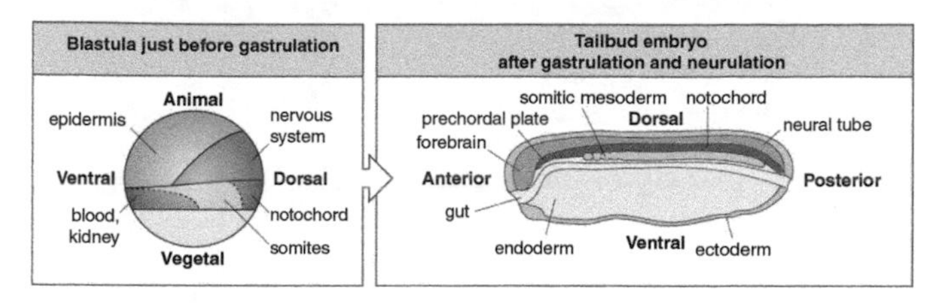

3. Fate map of the frog blastula showing the structures that will be formed in the tailbud embryo. After gastrulation and neurulation, the embryo has elongated and begins to take on the shape of a tadpole

such as muscle, cartilage, and neurons differentiate. Within 48 hours, the embryo has become a feeding tadpole with typical vertebrate features.

At the blastula stage, it is possible to make a fate map of the embryo (Figure 3). This shows what various regions of the blastula will give rise to later on, and is made by labelling the cells at this stage and then following their later development. As can be seen, internal regions like the endoderm are on the outside of the embryo and will move inside during gastrulation. The fate of many of the cells is not fixed at this stage and, if moved to another region, cells may develop according to their new location. However, with time, their fate becomes determined. If the region of the gastrula that will normally give rise to an eye is grafted into the trunk region of a slightly later stage—the neurula—the graft forms structures typical of its new location, such as notochord and somites. If, however, the eye region from a neurula is grafted into the trunk region it develops as an eye-like structure, since at this later stage its fate has become determined. Early vertebrate embryos have considerable capacity for regulation when pieces are removed or are transplanted to a different part of the same embryo. This implies considerable developmental plasticity at this early stage and also that the actual committed fate of cells is heavily dependent on the signals they receive from neighbouring cells.

Chapter 1
Cells

Development results from the coordinated behaviour of cells, which is almost entirely determined by which proteins the cell contains. Cells in the embryo are small and are surrounded by a cell membrane that determines what molecules can get in or out. Inside the cell are a number of smaller membrane-bound structures such as mitochondria, which produce the energy for the cell, and the cell nucleus, which contains the chromosomes. The chromosomes contain the DNA that makes up the **genes** which code for proteins. We humans have about 25,000 genes.

Proteins are long strings of 20 different types of amino acid subunits. Their sequence determines the shape and function of the protein, for example as an enzyme or a muscle protein. Each DNA strand is also a string of four different types of subunits, known as nucleotides. DNA acts as a coding region for proteins: for each protein there is a stretch of DNA—a gene—that codes for the sequence of amino acids in a protein. The system is rather like the Morse code where dots and dashes code for each of the letters in the alphabet. The sequence of DNA nucleotides, read three at a time, corresponds to the sequence of the amino acids along the protein; each set of three nucleotides codes for one amino acid. When a gene is active, its DNA sequence is first transcribed into an intermediary molecule, messenger RNA (mRNA), and this is used as a template for synthesizing the protein, using a similar three-nucleotide code for each amino acid in the protein.

Whether a gene is transcribed into mRNA depends on the binding of special proteins—**transcription factors**—to particular control regions in the DNA (Figure 4). These control regions do not encode proteins but provide recognition sites for the transcription factors and for the protein machine (RNA polymerase) that transcribes the code from DNA to mRNA. Some control regions are present near the coding region, while others may be far away. Only if the correct control regions are occupied by the right transcription factors can a gene be transcribed. The gene remains active (switched on) so long as the control regions are activated. The importance of these control regions cannot be overemphasized. The transcription factor protein produced by one gene can activate (or even inactivate) several other genes and so a network of gene interactions is set up that determines cell behaviour and how it changes with time. Some genes do not code for proteins but instead encode microRNAs, small RNA molecules that interfere with the translation of specific mRNAs into protein.

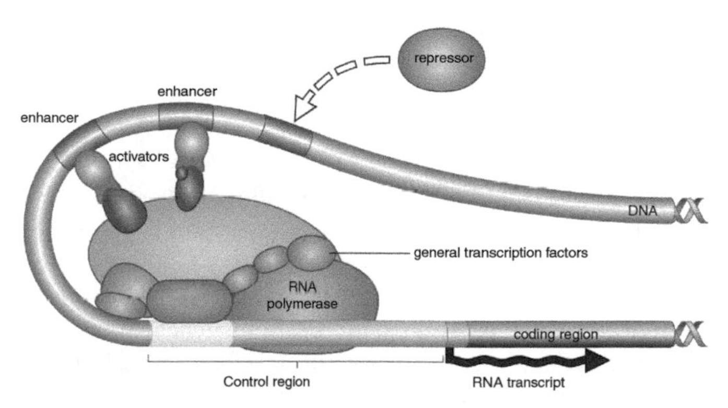

4. The transcription of a gene is carried out by RNA polymerase. This process is regulated by proteins (transcription factors) that bind to the control regions of gene which can be close to the coding region or at remote sites, such as the enhancers shown here

A change (mutation) in sequence of nucleotides in the DNA can occur in the coding region of a gene, and so alter the normal sequence of amino acids in the protein it encodes. This can alter the character of the protein, by altering how it folds or functions, and can result in a faulty protein being made, which can have serious positive or negative consequences on cell behaviour. Mutations that alter protein function in egg or sperm cells are the basis of evolution, as the mutation will be passed on to the next generation. Mutations in the DNA control regions can also affect cell behaviour as they determine when and in what cell a gene is made active and can be translated into a protein.

The major processes involved in development are: **pattern formation**; **morphogenesis** or change in form; cell differentiation by which different types of cell develop; and growth. These processes involve cell activities, which are determined by the proteins present in the cells. Genes control cell behaviour by controlling where and when proteins are synthesized, and cell behaviour provides the link between gene action and developmental processes. What a cell does is determined very largely by the proteins it contains. The hemoglobin in red blood cells enables them to transport oxygen; the cells lining the vertebrate gut secrete specialized digestive enzymes. These activities require specialized proteins that are not involved in the 'housekeeping' activities that are common to all cells and keep them alive and functioning. Housekeeping activities include the production of energy and the metabolic pathways involved in the breakdown and synthesis of molecules necessary for the life of the cell. In development we are concerned primarily with those proteins that make cells different from one another and make them carry out the activities required for development of the embryo. Developmental genes typically code for proteins involved in the regulation of cell behaviour.

All the information for embryonic development is contained within the fertilized egg. So how is this information interpreted to give rise to an embryo? Does the DNA contain a full description of the organism to which it will give rise; is it a blueprint for the organism? The answer is no. Instead, the fertilized egg contains a program of instructions for making the organism—a generative program—that determines where and when different proteins are synthesized and thus controls how cells behave. A descriptive program such as a blueprint or a plan describes an object in some detail, whereas a generative program describes how to make an object. For the same object, the programs are very different. Consider origami, the art of paper folding. By folding a piece of paper in various directions it is quite easy to make a paper hat or a bird from a single sheet. To describe in any detail the final form simply by marking regions on the flat piece of paper is really very difficult, and not of much help in explaining how to achieve it. Much more useful and easier to formulate are instructions on how to fold the paper. The reason for this is that simple instructions about folding have complex spatial consequences. In development, gene action similarly sets in motion a sequence of events that can bring about profound changes in the embryo. One can thus think of the genetic information in the fertilized egg as equivalent to the folding instructions in origami; both contain a generative program for making a particular structure.

Cells are, in a way, more complex than the embryo itself. There are thousands of different proteins and many copies of them in most cells in the embryo, and the network of interactions between proteins and DNA within any individual cell contains many more components and is very much more complex than the interactions between the cells of the developing embryo. However clever you think cells are, they are almost always far cleverer. Each of the basic cell activities involved in development, such as how to respond to external signals, to divide into two, or to move, is the result of interactions within a population of many different proteins whose composition varies over time and between different locations in the cell.

An intriguing question is how many genes out of the total genome are developmental genes—that is, genes specifically required for embryonic development. This is not easy to estimate. In the nematode worm, at least 50 specific genes are needed to specify a small reproductive structure known as the vulva. This is a very small number compared to the thousands of genes that are active at the same time; some of these are essential to development in that they are necessary for maintaining life, but they provide no or little information that influences the course of development. Some studies suggest that in an organism with 20,000 genes, about 10% of the genes may be directly involved in development.

A major goal of developmental biology is thus to understand how genes control embryonic development, and to do this one must first identify which genes, out of the many thousands in the organism, are critically and specifically involved in controlling development. The general starting point is to identify and create DNA mutations that alter development in some specific and informative way. Many developmental mutations have been produced by inducing random mutations in large numbers of model organisms, by chemical treatments or irradiation by X-rays, and then screening for mutants of developmental interest. Using modern genetic and bioinformatics techniques, many developmental genes have been identified. Direct DNA sequence comparison with known genes from model organisms has been very helpful in identifying human developmental genes. Twin studies are also useful. Despite having identical genes, identical twins can develop considerable differences due to effects in the womb and as they grow up, and these tend to become more evident with age.

The fate of a group of cells in the early embryo can be determined by signals from other cells. Few signals actually enter the cells. Most signals are transmitted through the space outside of cells (the extracellular space) in the form of proteins secreted by one cell and detected by another. Cells may interact directly with each other by means of molecules located on their surfaces. In both these cases, the signal is generally received by receptor proteins in the cell membrane and is subsequently relayed through other signalling proteins inside the cell to produce the cellular response, usually by turning genes on or off. This process is known as signal transduction. These pathways can be very complex. They can be likened to a Rube Goldberg cartoon in which a man has a relay mechanism for raising his umbrella when it rains: the rain first causes a prune to expand and so light a lighter which starts a fire which boils a kettle which whistles and so frightens a monkey who jumps onto a swing which cuts a cord releasing birds who, when they fly out, raise the umbrella. The complexity of the signal transduction pathway means that it can be altered as the cell develops so the same signal can have a different effect on different cells.

How a cell responds to a particular signal depends on its internal state and this state can reflect the cell's developmental history—cells have good memories. Thus, different cells can respond to the same signal in very different ways. So the same signal can be used again and again in the developing embryo. There are thus rather few signalling proteins.

Techniques from molecular biology and genetics have revolutionized the study of developmental biology over the past few decades. Approaches are also being used to identify all the genes involved in a particular developmental process. The identification of all the genes expressed in a particular tissue or at a particular stage of development can be accomplished by carrying out genome-wide screens for gene expression. This technology enables the amounts of mRNA transcripts of thousands of genes to be measured simultaneously. Other technical advances are the great improvement in computer-aided microscopic imaging techniques, and the development of fluorescent labels in a vast range of colours that allow the imaging of living embryos and for grafts to be followed.

Chapter 2
Vertebrates

All vertebrates, despite their many outward differences, have a similar basic body plan—the segmented backbone or vertebral column surrounding the spinal cord, with the brain at the head end enclosed in a bony or cartilaginous skull. These prominent structures mark the antero-posterior axis with the head at the anterior end. The vertebrate body also has a distinct dorso-ventral axis running from the back to the belly, with the spinal cord running along the dorsal side and the mouth defining the ventral side. The antero-posterior and dorso-ventral axes together define the left and right sides of the animal. Vertebrates have a general bilateral symmetry around the dorsal midline so that outwardly the right and left sides are mirror images of each other though some internal organs such as the heart and liver are arranged asymmetrically. How these axes are specified in the embryo is a key issue.

All vertebrate embryos pass through a broadly similar set of developmental stages and the differences are partly related to how and when the axes are set up, and how the embryo is nourished. Yolk provides all the nutrients for fish, amphibian, reptile, and bird embryonic development, and for the few egg-laying mammals such as the platypus. By contrast, the eggs of most mammals are small and non-yolky, and the embryo is nourished for the first few days by fluids in the mother. The mammalian embryo develops specialized external membranes that surround and protect the embryo, and through which it receives nourishment from the mother via the placenta.

After fertilization, the egg undergoes a number of cell divisions known as cleavage which, in the frog, forms the spherical blastula but in the chick or mammals the corresponding structure is not a hollow sphere, but a layer of cells called the **epiblast**. The epiblast, like the blastula, gives rise to the three layers of the embryo (ectoderm, mesoderm, and endoderm) during gastrulation.

This can be seen in chick development as the early chicken embryo develops as a flat disc of cells, the epiblast, overlying a massive yolk. The large yolky chick egg cell is fertilized and begins to undergo cell division while still in the hen's oviduct. During the 20-hour passage down the oviduct, the egg becomes surrounded by egg white, and the eggshell. At the time of laying, the embryo is composed of some 20,000–60,000 cells.

Mammalian eggs are much smaller than those of chicks or frogs, and they contain no yolk. The unfertilized egg is shed from the ovary into the oviduct and is surrounded by a protective external coat. Fertilization takes place in the oviduct and cleavage starts. There is no clear sign of axes in the mouse egg, and the highly regulative nature of early development in the mouse argues against the importance of maternal determinants. The early cleavages give rise to two distinct groups of cells—the trophectoderm and the inner cell mass (Figure 5). The trophectoderm will give rise to extra-embryonic structures such as the placenta, through which the embryo gains nourishment from the mother, while the embryo proper develops from the inner cell mass. The cells of the inner cell mass are **pluripotent**—they can give rise to all the cell types in the embryo. Inner cell mass cells can be isolated and grown in culture to produce pluripotent embryonic stem cells (ES cells) as discussed later.

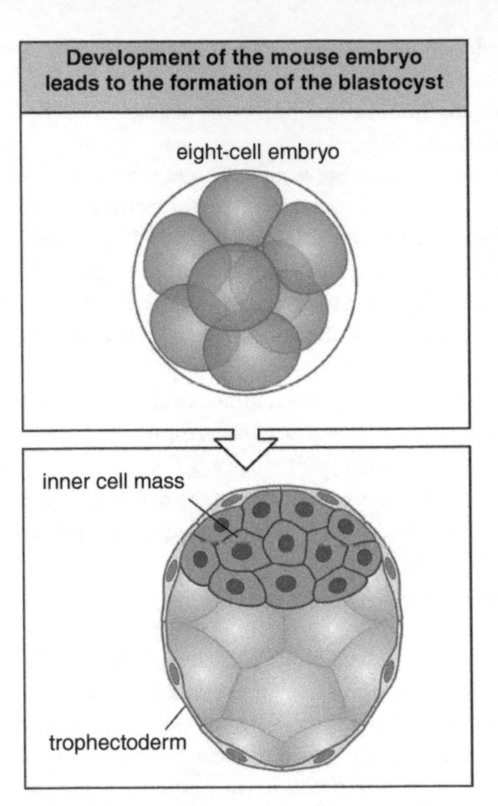

5. Cleavage of the mouse embryo leads to the inner cell mass lying inside an epidermal layer—the trophectoderm

A quite rare but nevertheless important event before gastrulation in mammalian embryos, including humans, is the splitting of the embryo into two, and identical twins can then develop. This shows the remarkable ability of the early embryo to regulate and develop normally when half the normal size, just like the Driesch experiment. It also makes clear that the early embryo should not be thought of as a human being as it can still develop into two people.

How are the antero-posterior and dorso-ventral axes established in vertebrate embryos and are they already present in the egg, or specified later? The establishment of axes occurs at the earliest stages in development of both the frog and zebrafish and is exclusively under the control of maternal factors present in the egg. The frog egg possesses a distinct axis even before it is fertilized. The upper region of the egg is the pigmented animal pole, whereas most of the yolk is located toward the opposite, unpigmented end, the vegetal pole. These differences define the animal–vegetal axis. The spherical symmetry about the animal–vegetal axis is broken when the egg is fertilized. Sperm entry sets in motion a series of events that defines the dorso-ventral axis of the gastrula with the dorsal side forming more or less opposite the sperm's entry point. The first signalling centre that develops in the dorsal–vegetal region in the frog blastula is known as the blastula organizer or Nieuwkoop centre, which sets up the initial dorso-ventral polarity in the blastula. In the chick, the antero-posterior axis is determined by gravity when the early embryo rotates during its passage in the hen's uterus before it is laid. In mammals, there is no sign of axes or polarity in the fertilized egg or during early development, and it only occurs later by an as yet unknown mechanism.

The first indication of the antero-posterior axis of the chick embryo is a crescent-shaped ridge of small cells at the posterior end of the epiblast where specific genes are activated, and this defines where the primitive streak will form. The streak is first visible as a denser region that then gradually extends anteriorly as a narrow furrow to just over half way across the epiblast. During gastrulation, epiblast cells converge on the primitive streak and move through the furrow and inward and then spread out anteriorly and laterally beneath the upper layer (Figure 6). The cells that move down through the streak give rise to mesoderm and endoderm, whereas the cells that remain on the surface of the epiblast give rise to the ectoderm. At the anterior end of the streak is a cluster of cells known as Hensen's node. This is the major organizing centre for the early chick embryo, equivalent to the Spemann organizer in amphibians, and it can induce a new primitive streak if transplanted to another early embryo. The epiblast can be cut up into four regions with two diagonal cuts and each will form a streak and give rise to a normal embryo—impressive regulation.

After the primitive streak has elongated to its full length it begins to regress, with Hensen's node moving back towards the posterior end of the embryo. As the node regresses, the notochord is laid down in its wake and the mesoderm immediately on each side of the notochord begins to form the somites (Figure 7). In the chick, the first pair of somites is formed at about 24 hours after laying and new ones are formed at intervals of 90 minutes. Later, they will form vertebrae and they are also the source of body muscle. As the notochord forms, the neural tube that will form the brain and spinal cord develops above it in a manner similar to the frog (Figure 8).

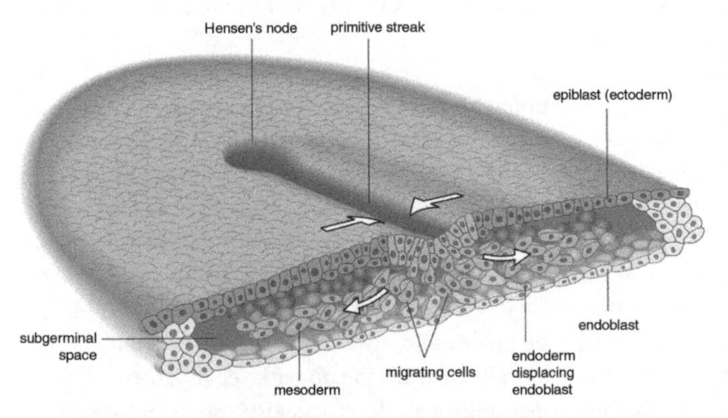

6. Gastrulation in the chick embryo. Cells in the epiblast converge on the primitive streak and migrate through it, giving rise to endoderm and mesoderm. The remaining cells of the epiblast give rise to the ectoderm

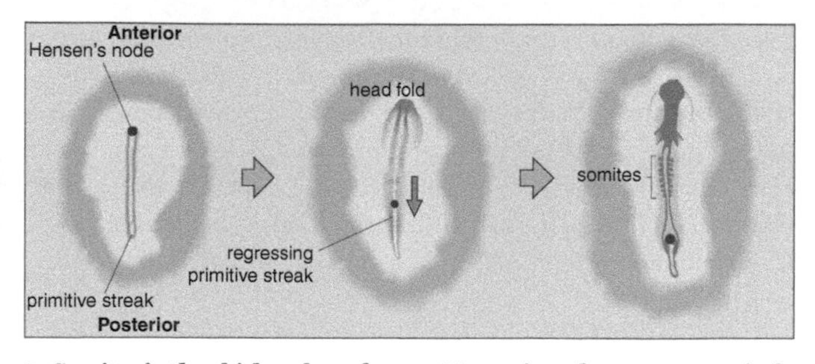

7. Somites in the chick embryo form as Hensen's node moves posteriorly

Gastrulation is followed by the formation of the neural tube, which is the early embryonic precursor of the central nervous system. The earliest visible sign is formation of the neural folds, which form on the edges of the neural plate, an area of ectoderm cells overlying the notochord. The folds rise up, fold toward the midline, and fuse together to form the neural tube, which sinks beneath the epidermis. Neural crest cells detach from the top of the neural tube on either side of the site of

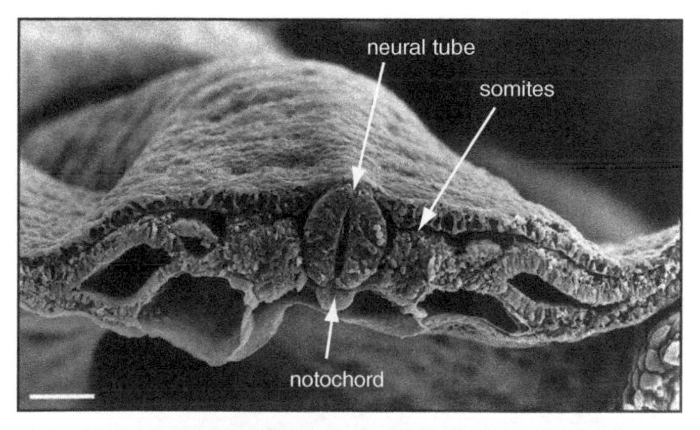

8. Early somites and neural tube in the chick embryo. The neural tube (centre) has somites on either side and the notochord beneath it

fusion and migrate throughout the body to form various structures, as we will see later. The anterior neural tube gives rise to the brain; further back, the neural tube overlying the notochord will develop into the spinal cord. The embryo now resembles a tadpole and we can recognize the main vertebrate features. At the anterior end, the brain is already divided up into a number of regions, and the eyes and ears have begun to develop.

How is left–right established? Vertebrates are bilaterally symmetric about the midline of the body for many structures, such as eyes, ears, and limbs, but most internal organs are asymmetric. In mice and humans, for example, the heart is on the left side, the right lung has more lobes than the left, the stomach and spleen lie towards the left, and the bulk of the liver is towards the right. This handedness of organs is remarkably consistent, but there are rare individuals, about one in 10,000 in humans, who have the condition known as *situs inversus*, a complete mirror-image reversal of handedness. Such people are generally asymptomatic, even though all their organs are reversed.

Specification of left and right is fundamentally different from specifying the other axes of the embryo, as left and right have meaning only after the antero-posterior and dorso-ventral axes have been established. If one of these axes were reversed, then so too would be the left–right axis and this is the reason that handedness is reversed when you look in a mirror—your dorso-ventral axis is reversed, and so left becomes right and vice versa. The mechanisms by which left–right symmetry is initially broken are still not fully understood, but the subsequent cascade of events that leads to organ asymmetry is better understood. The 'leftward' flow of extracellular fluid across the embryonic midline by a population of ciliated cells has been shown to be critical in mouse embryos in inducing asymmetric expression of genes involved in establishing left versus right.

The antero-posterior patterning of the mesoderm is most clearly seen in the differences in the somites that form vertebrae: each individual vertebra has well defined anatomical characteristics depending on its location along the axis. The most anterior vertebrae are specialized for attachment and articulation of the skull, the vertebrae of the neck are followed by the rib-bearing vertebrae and then those of the lumbar region, which do not bear ribs, and, finally, those of the sacral and caudal regions. Patterning of the skeleton along the body axis is based on the somite cells acquiring a positional value that reflects their position along the axis and so determines their subsequent development.

Somites are formed in a well-defined order along the antero-posterior axis and give rise to: the bone and cartilage of the trunk, including the spinal column; the skeletal muscles; and the dermis of the skin on the dorsal side of the body. The vertebrae, for example, have characteristic shapes at different positions along the spine. Somites are formed in pairs, one on either side of the notochord. Somite formation is largely determined by an internal 'clock' in the pre-somite mesoderm. This clock is represented by periodic cycles of gene expression, in the chick embryo, whose expression sweeps from the posterior to the anterior with a period of 90 minutes.

It is the **Hox genes** that define positional identity along the antero-posterior axis and they were originally identified in the fly, as will be seen. The Hox genes are members of the large family of homeobox genes that are involved in many aspects of development and are the most striking example of a widespread conservation of developmental genes in animals. The name homeobox comes from their ability to bring about a homeotic transformation, converting one region into another. Most vertebrates have clusters of Hox genes on four different chromosomes. A very special feature of Hox gene expression in both insects and vertebrates is that the genes in the clusters are expressed in the developing embryo in a temporal and spatial order that reflects their order on the chromosome. Genes at one end of the cluster are expressed in the head region, while those at the other end are expressed in the tail region. This is a unique feature in development, as it is the only known case where a spatial arrangement of genes on a chromosome corresponds to a spatial pattern in the embryo. The Hox genes provide the somites and adjacent mesoderm with positional values that determine their subsequent development. Morphological changes occur if their pattern of expression is altered. Mice in which the *Hoxd3* gene is deleted show structural defects in the first and second vertebra, where this gene is normally strongly expressed.

It is now possible to identify all the genes involved in a developmental process. In order to understand how each of these genes controls development, strains of mice with a particular mutant gene that affects development can now be produced relatively routinely and animals that have an additional or altered gene are known as transgenic animals. Two main techniques for producing transgenic mice are currently in use. One is to inject the DNA encoding the required gene and any necessary regulatory regions directly into the male nucleus of a newly fertilized egg. A newer technique for producing transgenic mice uses cells removed from the inner cell mass of an early mouse and then mutated in culture. As we shall see later, these cells are pluripotent and known as embryonic stem cells (ES cells). ES cells injected into the cavity of an early mouse embryo become incorporated into the inner cell mass and so will become part of all the embryo's tissues, even giving rise to germ cells. To generate transgenic mice with a particular mutation, the ES cells are mutated while growing in culture and before they are introduced into an early mouse embryo. Because it is not possible to generate such changes in frog and chick embryos, a different technique known as gene silencing is particularly useful. Morpholino antisense RNAs are designed to be complementary to a specific mRNA. When injected into the cells of an embryo, they bind to just the target mRNA, preventing its translation into protein.

Chapter 3
Invertebrates and plants

Drosophila

The advances made in understanding development of the fruit fly *Drosophila melanogaster* have had a major impact on our understanding of development in other organisms including vertebrates. We are much more like flies in our development than you might think. Many of the genes that control the development of flies are similar to those controlling development in vertebrates, and indeed in many other animals. It seems that once evolution finds a satisfactory way of developing animal bodies, it tends to use the same mechanisms and molecules over and over again with, of course, some important modifications.

Many of the developmental mutations that have led to our present understanding of early fly development, and provided key insights into development, came from a brilliantly successful screening program that searched the fly genome systematically for mutations affecting the patterning of the early embryo. Its success was recognized with the award of a Nobel Prize in 1995.

After fertilization and fusion of the sperm and egg nuclei, the fused nucleus undergoes a series of rapid duplications and divisions, one about every 9 minutes, but, unlike in most embryos, there is initially no cleavage of the cytoplasm and no formation of cell membranes to separate the nuclei. The result after 12 nuclear divisions is around 6000 nuclei present in a layer beneath the cell membrane, the embryo essentially remaining a single cell with many nuclei. There is early pattern formation at this stage and, shortly afterwards, membranes grow in from the surface to enclose the nuclei and form a single layer of cells. All the future tissues, except for the germline cells, are derived from this single layer of cells.

The insect body is bilaterally symmetrical and has two distinct and largely independent axes: the antero-posterior and dorso-ventral axes, which are at right angles to each other. These axes are already partly set up in the fly egg, and become fully established and patterned in the very early embryo. Along the antero-posterior axis the embryo becomes divided into a number of segments, which will become the head, thorax, and abdomen of the larva. A series of evenly spaced grooves forms more or less simultaneously and these demarcate parasegments, which later give rise to the segments of the larva and adult. Of the fourteen larval parasegments, three contribute to mouthparts of the head, three to the thoracic region, and eight to the abdomen. The fly larva has neither wings nor legs; these and other organs are formed when the larva later undergoes hormone-induced metamorphosis to its adult form, as will be described later. These structures are, however, already present in the larva as imaginal discs, small sheets of cells containing about 40 cells each at the time they are formed.

Development is initiated by a gradient of the protein Bicoid, along the axis running from anterior to posterior in the egg; this provides the positional information required for further patterning along this axis. Bicoid is a transcription factor and acts as a **morphogen**—a graded concentration of a molecule that switches on particular genes at different threshold concentrations, thereby initiating a new pattern of gene expression along the axis. Bicoid activates anterior expression of the gene *hunchback* (Figure 9). The *hunchback* gene is switched on only when Bicoid

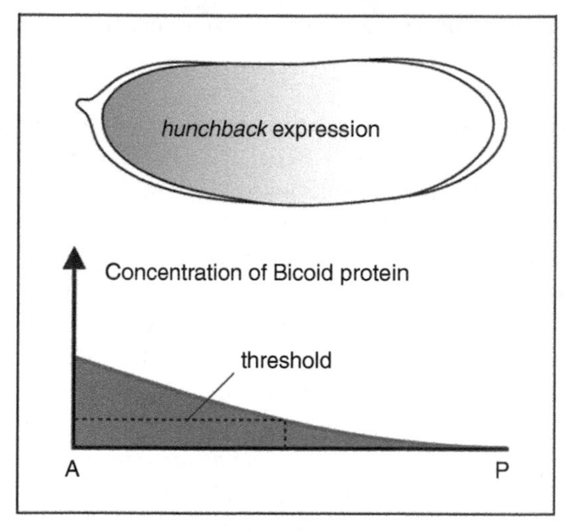

9. The gradient in the maternal Bicoid protein turns on the gene *hunchback* at a concentration above a threshold level

is present above a certain threshold concentration. The protein of the *hunchback* gene, in turn, is instrumental in switching on the expression of the other genes, along the antero-posterior axis.

The dorso-ventral axis is specified by a different set of maternal genes from those that specify the anterior-posterior axis, but by a similar mechanism. The initial dorso-ventral organization of the embryo is established at right angles to the antero-posterior axis, and the embryo initially becomes divided into four regions along the dorso-ventral axis, and this patterning is controlled by the distribution of the maternal protein Dorsal. Dorsal is graded along the ventral to dorsal axis and its effects on gene expression divide the dorso-ventral axis into well-defined regions. In the ventral-most region, where concentrations of Dorsal protein in the nucleus are highest, gastrulation results in a ventral band of prospective mesoderm cells moving into the interior of the embryo.

As mentioned above, the embryo becomes divided along the antero-posterior axis into a number of segments, parasegments, that are the fundamental units in the segmentation of the fly embryo. Once each parasegment is delimited, it behaves as an independent developmental unit, under the control of a particular set of genes. The parasegments are initially similar but each will soon acquire its own unique identity mainly due to Hox genes. How are the parasegments specified? Remarkably they are defined by the action of the pair-rule genes, each of which is expressed in a series of seven transverse stripes along the embryo, corresponding to every second parasegment. When pair-rule gene expression is visualized by staining for the pair-rule proteins, a striking zebra-striped embryo is revealed. At first sight, this type of patterning would seem to require some underlying periodic process, such as the setting up of a wave-like concentration of a chemical, with each stripe forming at the crest of a wave. It was surprising, therefore, to discover that each stripe is specified independently by the pattern of proteins set up earlier. Figure 10 is an example of how the pair-rule gene specifying the third stripe, *even-skipped* 2, is specified. The pair-rule genes thus require complex control regions with multiple binding sites for each of the different factors. Examination of the control regions of pair-rule genes reveals seven separate regions, each controlling the localization of a different stripe. This is an excellent model for how complex pattern can develop from the organization of control regions and the proteins that bind to them.

The parasegments give rise to the segments of the larva and later to those of the adult fly. The epidermis in each segment not only becomes patterned into bands of epidermal cells of different types, but each cell acquires an individual antero-posterior polarity, reflected by the fact that the hairs and bristles on the adult fly

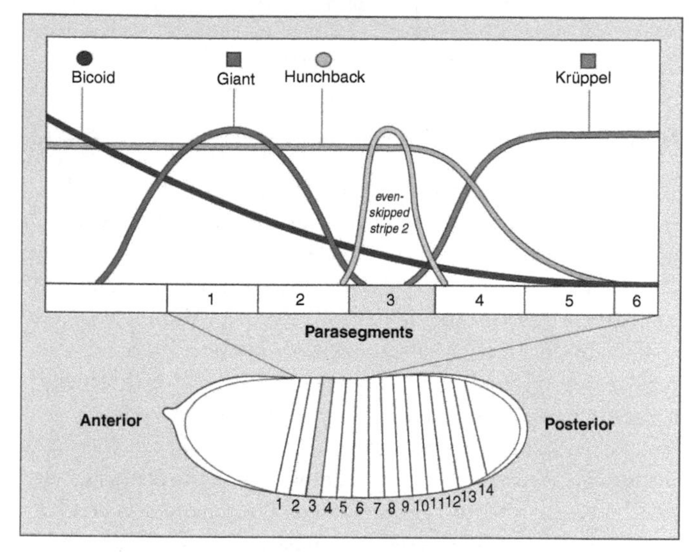

10. The specification of the second *even-skipped* stripe as a narrow stripe in parasegment 3. The proteins Bicoid and Hunchback activate the *even-skipped* gene, and Giant and Krüppel repress it

abdomen all point backwards. This type of cell polarity is called planar cell polarity.

Each segment in the fly has a unique identity, most easily seen in the larva in the characteristic pattern of denticles (pointed projections) on the surface. What makes the segments different from each other? Their identity is specified by the Hox genes, which, as we have seen, provide positional identity in vertebrates, but were first identified in the fly. The first evidence for the existence of these genes that specify segment identity came from unusual and striking mutations in the fly that produced homeotic transformations—the conversion of one segment into another, such as an antenna into a leg. In the fly, Hox genes are on just one chromosome and, as described for vertebrates, their order of expression along the antero-posterior axis corresponds with their order along the chromosome. The site of appendages such as legs along the body is determined by the Hox genes.

Nematodes

The early embryos of many invertebrates contain far fewer cells than those of flies and vertebrates, with each cell acquiring a unique identity at an early stage of development. In the nematode, for example, there are only 28 cells when gastrulation starts, compared with thousands in the fly. There is an old, and now less fashionable, distinction sometimes made between so-called regulative and mosaic development—the former involving mainly cell–cell interactions, whereas the latter is based on localized factors in cells and their asymmetric distribution at cell division to the two daughter cells. In mosaic development, the factors are in localized regions of the egg. A feature of the nematode is that cell fate is often specified on a cell-by-cell basis, a typical characteristic of mosaic development, and in general does not rely on positional information established by gradients of morphogens. Specification on a cell-by-cell basis often makes use of asymmetric cell division and the unequal distribution of cytoplasmic factors (Figure 11). However, asymmetric cell division in the early stages of development does not mean that cell–cell interactions are absent or unimportant in these organisms.

The free-living soil nematode *Caenorhabditis elegans* is an important model organism in developmental biology. Its advantages are its suitability for genetic analysis, the small number of cells and their fixed lineage, and the transparency of the embryo, which allows the formation of each cell to be observed. Its study led to key discoveries concerning genetic control of organ development and programmed cell death (apoptosis), and was recognized by a Nobel Prize.

It is a triumph of direct observation that the complete lineage of every cell in the nematode has been worked out. The pattern of cell

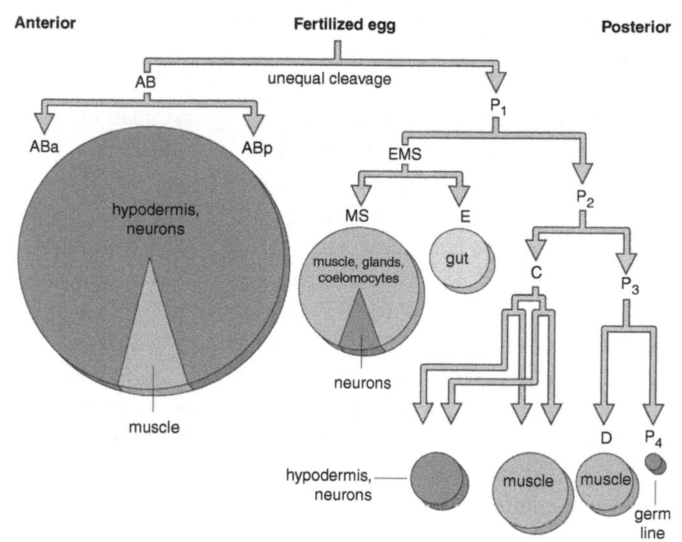

11. Cell lineage of the early nematode *Caenorhabditis elegans* embryo. The hypodermis is part of the outer layer

division is largely invariant—it is virtually the same in every embryo. The larva, when it hatches, is made up of 558 cells, and after four further moults this number has increased to 959, excluding the germ cells which vary in number. This is not the total number of cells derived from the egg, as 131 cells undergo programmed cell death or apoptosis (cell suicide) during development, which is discussed in more detail later. As the fate of every cell at each stage is known, a fate map can be accurately drawn at any stage and thus has a precision not found in any vertebrate. As with any fate map, however, this precision in no way implies that the lineage must determine the fate or that the fate of the cells cannot be altered. As we shall see, interactions between cells have a major role in determining cell fate in the nematode.

Around 1700 genes have been identified as affecting development, and many nematode developmental genes are related to genes that control development in the fly and other animals and they include the Hox genes and genes for signalling proteins.

Before fertilization, there is no evidence of any asymmetry in the nematode egg, but sperm entry sets up an antero-posterior polarity in the fertilized egg that determines the position of the first cleavage division. The first cleavage is asymmetric and defines the antero-posterior axis. It generates an anterior AB cell and a smaller posterior P1 cell. The existence of polarity in the fertilized egg becomes evident before the first cleavage. A cap of microfilaments forms at the future anterior end, and a set of granules—the P granules which contain maternal mRNAs and proteins required for development of the germline cells—become localized at the future posterior end where P1 will develop.

Cell differentiation in the nematode is closely linked to the pattern of cell division. Each cell undergoes a unique and nearly invariant series of cleavages that successively divide cells into anterior and posterior daughter cells. Cell fate appears to be specified by whether the final differentiated cell is descended through the anterior or posterior cell at each division. Despite the highly determinate cell lineage in the nematode, cell–cell interactions are involved in specifying the dorso-ventral axis.

The timing of events in development is not well understood, so an example in nematode development is of significance. It has been found that timing in nematode development is under genetic control that involves microRNAs, short RNA molecules that do not code for proteins, but alter the expression of other mRNAs. Embryonic development gives rise to a larva of 558 cells and there are then four larval stages before an adult is produced. Because each cell in the developing nematode can be identified by its lineage and position, genes that control the fates of individual cells at specific times in development can also be identified. Mutations that alter the timing of developmental events at larval stages were discovered and they illustrate the control of this process by microRNAs. Different mutations in such genes can produce either 'retarded' or 'precocious' development. Genes that control timing of developmental events may do so by controlling the concentration of some substance.

Plant development

Because plant cells have rigid cell walls and, unlike animal cells, cannot move, a plant's development is very much the result of patterns of oriented cell divisions and increase in cell size. Despite this difference, cell fate in plant development is largely determined by similar means as in animals—by a combination of positional signals and intercellular communication. As well as communicating by extracellular signals and cell-surface interactions, plant cells are interconnected by cytoplasmic channels known as plasmodesmata, which allow movement of proteins such as transcription factors directly from cell to cell.

The logic behind the spatial layouts of gene expression that pattern a developing flower is similar to that of Hox gene action in patterning the body axis in animals, but the genes involved are completely different. One general difference between plant and animal development is that most of the development occurs not in the embryo but in the growing plant. Unlike an animal embryo, the mature plant embryo inside a seed is not simply a smaller version of the organism it will become. All the 'adult' structures of the plant—shoots, roots, stalks, leaves, and flowers—are produced in the adult plant from localized groups of undifferentiated cells known as **meristems**.

Two meristems are established in the embryo, one at the tip of the root and the other at the tip of the shoot. These persist in the adult plant, and almost all the other meristems, such as those in developing leaves and flower shoots, are derived from them. Cells within meristems can divide repeatedly and can potentially give rise to all plant tissues and organs. Another important difference between plant and animal cells is that a complete, fertile plant can develop from a single differentiated somatic cell and not just from a fertilized egg. This suggests that, unlike the differentiated cells of adult animals, some differentiated cells of the adult plant may retain **totipotency** and so behave like animal embryonic stem cells.

The small cress-like weed *Arabidopsis thaliana* has become the model plant for genetic and developmental studies. It has just two sets of chromosomes, which contain about 27,000 protein-coding genes. It is an annual, flowering in the first year of growth, and develops as a small ground-hugging rosette of leaves, from which a branched flowering stem is produced with a flowerhead, or inflorescence, at the end of each branch. It develops rapidly, with a total life cycle in laboratory conditions of some 6–8 weeks, and, like all flowering plants, mutant strains and lines can easily be stored in large quantities in the form of seeds.

Following fertilization, the embryo develops inside the ovule, the structure that gives rise to and contains the female reproductive cells within the flower, taking about 2 weeks to form a mature seed which is shed from the plant. The seed will remain dormant until suitable external conditions trigger germination. At germination, the shoot and root elongate and emerge from the seed. Once the shoot emerges above ground it starts to photosynthesize (use the energy from sunlight to make carbon compounds from carbon dioxide) and forms the first true leaves at the shoot apex. About four days after germination, the seedling is a self-supporting plant. Flower buds are usually visible on the young plant 3–4 weeks after germination and will open within a week.

During embryogenesis, the shoot–root polarity of the plant body, which is known as the apical–basal axis, is established, and the shoot and root meristems are formed. Development of the *Arabidopsis* embryo involves a rather invariant pattern of cell division. The first division is at right angles to the long axis, dividing it into an apical cell and a basal cell and establishing an initial polarity that is carried over into the apical–basal axis of the plant. The next divisions produce an embryo of around 32 cells (Figure 12). The embryo elongates and the cotyledons (seed leaves) start to develop as wing-like structures at one end, while an embryonic root forms at the other. This stage is known as the heart stage. Apical meristems capable of continued division are located at each end of this axis: the one lying between the cotyledons gives rise to the shoot, while the one at the opposite end of the axis will form the root. The region in between the embryonic root and the future shoot will become the seedling stem, also known as the hypocotyl. Almost all adult plant structures are derived from the apical meristems.

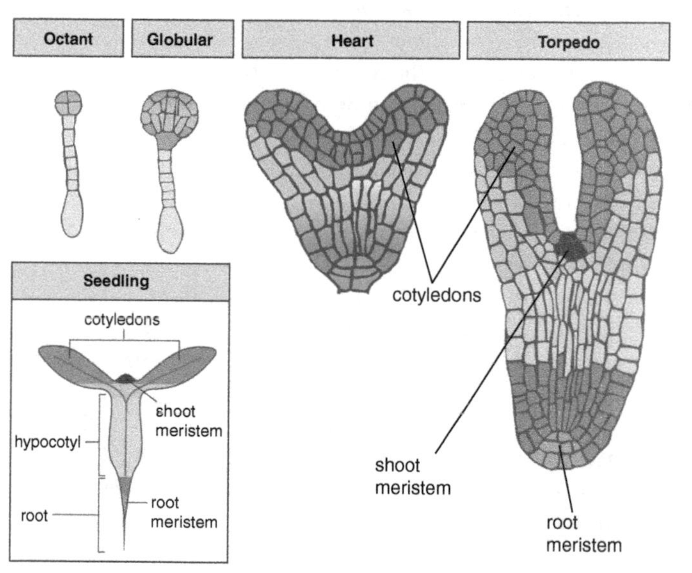

12. Fate map of the *Arabidopsis thaliana* embryo which will form a seedling (inset)

The small organic molecule auxin is one of the most important and ubiquitous chemical signals in plant development and plant growth. It causes changes in gene expression. In some cases, auxin appears to be acting as a classical morphogen, forming a concentration gradient and specifying different fates according to a cell's position along the gradient. The earliest known function of auxin in *Arabidopsis* is in the very first stage of embryogenesis, where it establishes the apical–basal axis. Immediately after the first division, auxin is actively transported from the basal cell into the apical cell, where it accumulates. The auxin is required to specify the apical cell, which gives rise to the shoot apical meristem. Through the subsequent cell divisions, transport of auxin continues until the embryo has about 32 cells. The apical cells of the embryo then start to produce auxin, and the direction of auxin transport is suddenly reversed.

A meristem contains a small central zone of self-renewing stem cells. Meristem stem cells are maintained in the self-renewing state by cells underlying the central zone that form the organizing centre. It is the microenvironment maintained by the organizing centre that gives stem cells their identity. Cells of the organizing centre express the protein Wuschel, a homeobox transcription factor which is required to produce a signal that gives the overlying cells their stem-cell identity. Cells leave the periphery of the meristem to form organs such as leaves or flowers, and are replaced from a small central zone of slowly dividing, self-renewing stem cells at the tip of the meristem. The stem cells behave in the same way as animal stem cells and they can divide to give one daughter that remains a stem cell and one that gives rise to plant tissue. This stem cell continues to divide and its descendants are displaced towards the peripheral zone of the meristem, where they become founder cells for a new organ, and leave the meristem, and differentiate.

Most of the *Arabidopsis* embryonic apical meristem gives rise to the first six leaves, whereas the remainder of the shoot, including all the flowerheads, is derived from a very small number of embryonic cells at the centre of the meristem. Leaves develop from groups of founder cells within the peripheral zone of the shoot apical meristem. In the structure that will become the leaf, two new axes are established that relate to the future leaf: the proximo-distal axis (leaf base to leaf tip) and the upper surface to lower surface. As the shoot grows, leaves are generated within the meristem at regular intervals and with a particular spacing. Leaves are arranged along a shoot in a variety of ways in different plants, and the particular arrangement is known as phyllotaxis. A common arrangement is the positioning of single leaves spirally up the stem, which can sometimes form a striking helical pattern in the shoot apex. In plants in which leaves are borne spirally, a new leaf primordium forms at the centre of the first available space outside the central region of the meristem and above the previous primordium. This pattern suggests a mechanism for leaf arrangement based on lateral inhibition in which each leaf primordium inhibits the formation of a new leaf within a given distance.

In the root meristem, the cells are organized rather differently from those in the shoot meristem, and there is a much more stereotyped pattern of cell division. The root meristem, like that of the shoot, is composed of an organizing centre, called the quiescent centre in roots, in which the cells divide only very rarely, and which is surrounded by stem-cell-like cells that give rise to the root tissue. The quiescent centre is essential for meristem function. Auxin plays a key role in patterning the growing root, and there is a stable auxin maximum concentration at the quiescent centre.

Flower development is dealt with in the chapter on organ formation.

Chapter 4
Morphogenesis

All animal embryos undergo a dramatic change in shape during their early development. This occurs principally during gastrulation, the process that transforms a two-dimensional sheet of cells into the complex three-dimensional animal body, and involves extensive rearrangements of cell layers and the directed movement of cells from one location to another. If pattern formation can be likened to painting, morphogenesis is more akin to modelling a formless lump of clay into a recognizable shape.

Change in form is largely a problem in cell mechanics and requires forces to bring about changes in cell shape and cell migration. Two key cellular properties involved in changes in animal embryonic form are cell contraction and cell adhesiveness. Contraction in one part of a cell can change the cell's shape. Changes in cell shape are generated by forces produced by the cytoskeleton, an internal protein framework of filaments. Animal cells stick to one another, and to the external support tissue that surrounds them (the extracellular matrix), through interactions involving cell-surface proteins. Changes in the adhesion proteins at the cell surface can therefore determine the strength of cell–cell adhesion and its specificity. These adhesive interactions affect the surface tension at the cell membrane, a property that contributes to the mechanics of the cell behaviour. Cells can also migrate, with contraction again playing a key role. An additional force that operates during morphogenesis, particularly in plants but also in a few aspects of animal embryogenesis, is hydrostatic pressure, which causes cells to expand. In plants there is no cell movement or change in shape, and changes in form are generated by oriented cell division and cell expansion. Cell division also plays a key role in animal changes in form.

Localized contraction can change the shape of the cells as well as the sheet they are in. For example, folding of a cell sheet—a very common feature in embryonic development—is caused by localized changes in cell shape (Figure 13). Contraction on one side of a cell results in it acquiring a wedge-like form; when this occurs among a few cells locally in a sheet, a bend occurs at the site, deforming the sheet. The localized cellular contraction is generated by protein filaments similar to, but simpler than, those in muscle. Changing the contacts between cells can also bring about a change in overall shape and can allow groups of cells to separate.

Many embryonic cells, like neural crest cells, can migrate over quite long distances. They move by extending a thin sheet-like layer of cytoplasm or long fine processes called filopodia, which attach to the surface over which they are moving Both these temporary structures are pushed outwards from the cell by the assembly of cytoskeletal filaments. Contraction of the muscle-like network at either the front or the rear of the cell then moves the cell forward.

The integrity of tissues in the embryo is maintained by adhesive interactions between cells and between cells and the extracellular matrix; differences in cell adhesiveness also help maintain the boundaries between different tissues and structures. Cells stick to each other by means of cell adhesion molecules, such as cadherins, which are proteins on the cell surface that can bind strongly to proteins on other cell surfaces. About 30 different types of cadherins have been identified in vertebrates. Cadherins bind to

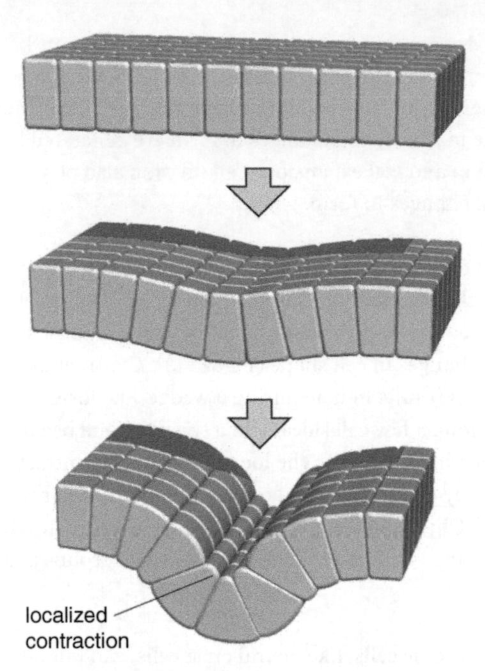

13. Localized contraction of a cell in a sheet can cause a change in the shape of the sheet, causing it to fold

each other; in general, a cadherin binds only to another cadherin of the same type, but they can also bind to some other molecules. Adhesion of a cell to the extracellular matrix, which contains proteins such as collagen, is by the binding of integrins in the cell membrane to these matrix molecules.

The particular adhesion molecules expressed by a cell determine which cells it can adhere to, and changes in the adhesion molecules expressed are involved in many developmental phenomena. Differences in cell adhesiveness can be illustrated by experiments in which cells from two different tissues are dissociated, mixed together, and then allowed to reaggregate. When cells of amphibian presumptive epidermis and presumptive neural plate are dissociated, mixed, and left to reaggregate, they sort out to re-form the two different tissues (Figure 14). The epidermal cells are eventually found on the outer face of the aggregate, surrounding a mass of neural cells—the same types of cell are now in contact with each other. Mixed ectodermal and mesodermal cells similarly sort themselves out to form a mass of cells with this time the ectoderm on the outside and mesoderm on the inside. This sorting out is the combined result of cell movement and differences in adhesiveness. Initially, cells move about randomly in the mixed aggregate, exchanging weaker for stronger adhesions. The adhesive interactions between cells produce different degrees of surface tension that is sufficient to generate the sorting-out behaviour, just as two immiscible liquids such as oil and water separate out when mixed.

The first change in shape in animal embryonic development is the division of the fertilized egg by cleavage into a number of smaller cells leading in many animals to the formation of a hollow sphere of cells—the blastula composed of an epithelial sheet enclosing a fluid-filled interior (Figure 15). The development of such a structure from a fertilized egg depends both on particular patterns of cleavage and on changes in the way cells pack together, as shown in schematic form in the figure. Early cleavage patterns can vary widely between different groups of animals. In radial cleavage, cleavages occur at right angles to the egg surface and the first few cleavages produce tiers of cells that sit directly over each other. This type of cleavage is characteristic of sea urchins and vertebrates. The eggs of molluscs (e.g. snails) and annelids (e.g. earthworms) illustrate another cleavage pattern, called spiral cleavage, in which successive divisions are at planes at slight angles to each other, producing a spiral arrangement of cells. The amount of yolk in the egg can influence the pattern of cleavage. In yolky eggs undergoing otherwise symmetric cleavage, a cleavage furrow starts in the least yolky region and gradually spreads across the egg.

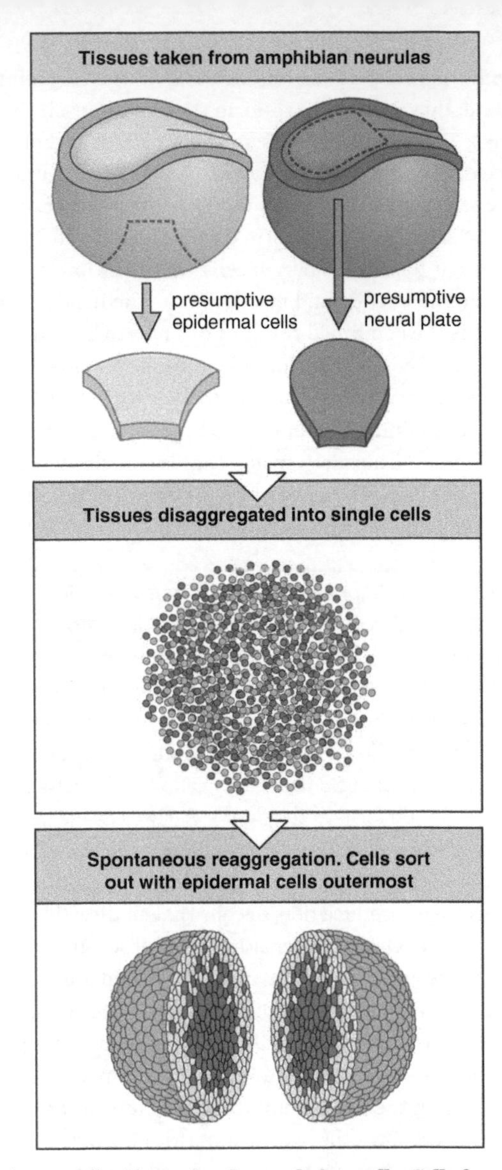

14. Sorting out of epidermal and neural plate cells. Cells from two regions of a frog embryo were separated into single cells and then allowed to reaggregate. The epidermal cells are all together on the outside with the neural cells inside

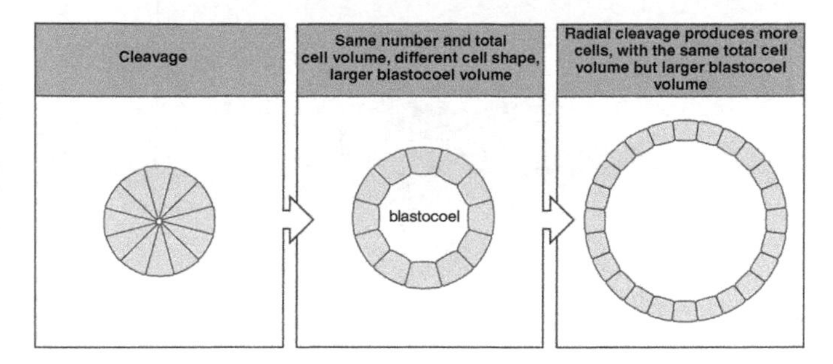

15. Cell cleavage and packing can determine the volume of the blastula

Gastrulation involves dramatic changes in the overall structure of the embryo, converting it into a complex three-dimensional structure. During gastrulation, a program of cell activity involving cell migration and changes in cell shape and adhesiveness remodels the embryo, so that the future endoderm and mesoderm move inside and only the ectoderm remains on the outside. The primary force for gastrulation is provided by changes in cell shape. The process is quite complex in vertebrates and is most easily seen in the sea urchin embryo, which also has the advantage of being transparent so that movies could be made of gastrulation. Cell cleavage after fertilization results in a spherical single-layered sheet of cells, surrounding a fluid-filled interior. The future mesoderm and endoderm are already specified and occupy a small region of the sphere, and the rest gives rise to ectoderm. Gastrulation begins with the mesodermal cells becoming motile; they become detached from each other and migrate inside as single cells to form a characteristic pattern on the inner surface of the sheet (Figure 16). They move by means of fine filopodia, which can be up to 40 micrometres long and can extend in several directions. When filopodia make contact with, and adhere to, the wall, they retract, drawing the cell body toward the point of contact. Each cell extends several filopodia, and so there can be competition between the filopodia, the cell being drawn towards that region of the wall where the filopodia make the most stable

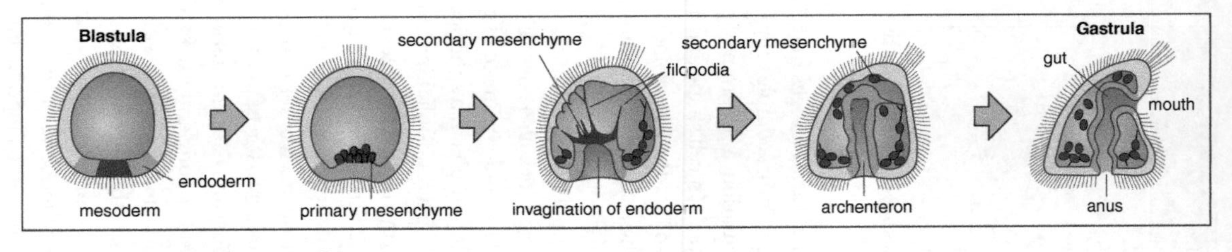

16. Gastrulation in the sea urchin

contact. The cells finally accumulate in the regions where the most stable contacts are made.

The entry of the mesoderm is followed by the invagination and extension of the endoderm to form the embryonic gut. The endoderm invaginates as a continuous sheet of cells. Formation of the gut occurs in two phases. During the initial phase, the endoderm invaginates to form a short, squat cylinder extending up to halfway across the interior. There is then a short pause before extension continues. In the second phase, the cells at the tip of the invaginating gut form long filopodia, which make contact with the wall. Their contractions pull the elongating gut until it comes in contact with and fuses with the mouth region, which forms a small invagination. This phase of gastrulation also involves convergent extension, due to active rearrangement of cells within the endodermal sheet.

Convergent extension plays a key role in gastrulation of other animals and other morphogenetic processes. It is a mechanism for elongating a sheet of cells in one direction while narrowing its width, and occurs by rearrangement of cells within the sheet, rather than by cell migration or cell division. It occurs, for example, in the extension of the mesoderm that elongates the antero-posterior axis in amphibian embryos. For convergent extension to take place, the axes along which the cells will intercalate and extend must already have been defined. The cells first become elongated in a direction at right angles to the antero-posterior axis—the medio-lateral direction (Figure 17). They also become aligned parallel to one another in a direction perpendicular to the direction of tissue extension. Active movement is largely confined to each end of these elongated bipolar cells, which shuffle in between each other—or intercalate—always along the medio-lateral axis, with some cells moving medially and some laterally. A mechanically similar process called radial intercalation causes the thinning of a multi-cell layer into a thinner sheet and its consequent extension around the edges, as seen during spreading of the ectoderm in the frog. Radial intercalation occurs in the multilayered ectoderm of the animal cap, in which cells intercalate in a direction perpendicular to the surface, moving from one layer into that immediately above. This leads to an increase in the surface area of the cell sheet, as well as its thinning.

Gastrulation in vertebrates involves a much more dramatic and complex rearrangement of tissues than in sea urchins, because of the need to produce a more complex body plan. In amphibians, fish, and birds, there is also the added complication of the presence of large amounts of yolk. But the outcome is the same: the transformation of a two-dimensional sheet of cells into a three-dimensional embryo, with ectoderm, mesoderm, and endoderm in the correct positions for further development of body structure. Gastrulation in mammals and birds occurs in the primitive streak and involves the convergence of epiblast cells on the midline, the separation of cells individually from the epiblast, and their internalization, followed by internal migration and convergent extension.

Gastrulation in the frog starts at a site on the dorsal side of the blastula, toward the vegetal pole. The first visible sign is the formation of bottle-shaped cells by some of the presumptive

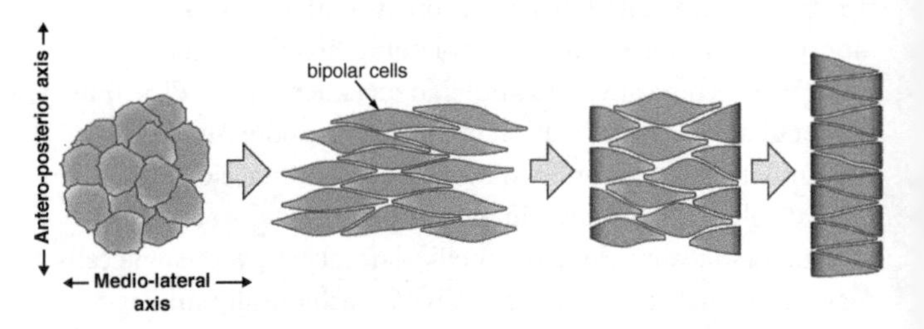

17. Convergent extension occurs by the movement of the bipolar cells, and the sheet narrows and extends

mesodermal cells. The formation of the bottle shape is due to apical constriction of the cell (at its top) and it forms a groove in the blastula surface, the blastopore, whose dorsal lip is the site of the Spemann organizer. The layer of mesoderm and endoderm starts to move in around the blastopore, and their movements and organization are complex. Convergent extension occurs in both the mesoderm and endoderm as they move in and, together with the elongation of the notochord, all these processes elongate the embryo in an antero-posterior direction.

In the chick, the mouse, and in humans, gastrulation is a little easier to describe and occurs through the primitive streak as described earlier. The epithelial epiblast—the future ectoderm—gives rise to both the mesoderm and endoderm as well. Epiblast cells become specified as mesodermal and endodermal cells in the streak and they leave the epiblast and move through the streak into the interior and form the gut and mesodermal tissues like muscle and cartilage and the blood supply as described earlier for the chick.

Neurulation in vertebrates results in the formation of the neural tube—a tube of epithelium derived from the dorsal ectoderm—which develops into the brain and spinal cord. Following induction by the mesoderm during gastrulation, the ectodermal cells that will give rise to the neural tube initially appear as a thickened plate of tissue—the neural plate—in which the cells have become more columnar. The vertebrate neural tube is formed by two different mechanisms in different regions of the body. The anterior neural tube, which forms the brain and the anterior spinal cord, is formed by the folding of the neural plate into a tube. The edges of the neural plate become raised above the surface, forming two parallel neural folds that come together along the dorsal midline of the embryo and fuse at their edges to form the neural tube, which then separates from the adjacent ectoderm (Figure 18). The posterior neural tube, by contrast, develops from a solid rod of cells that develops an interior cavity

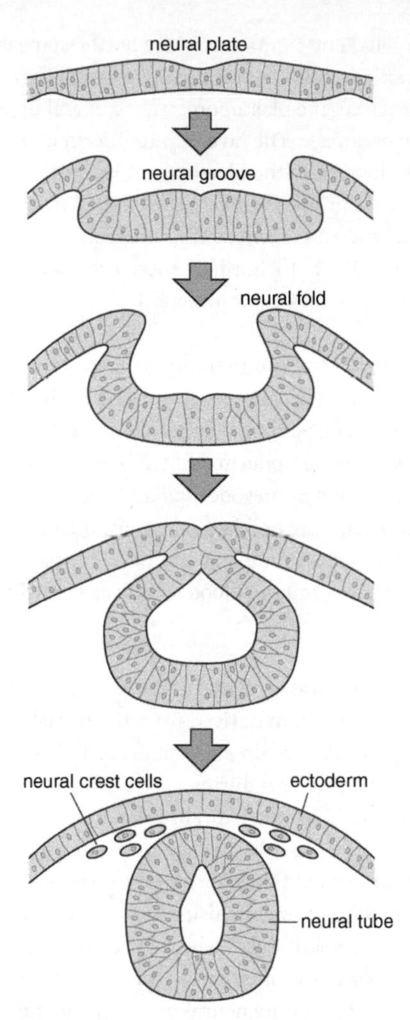

18. Neural tube formation

or lumen. The curvature of the neural plate and formation of the neural folds are due to changes in cell shape. The cells at the edge of the plate, where it is bending most, are constricted at their apical surface. Separation of the neural tube from the ectoderm after its formation is a result of changes in cell adhesiveness.

Neural crest cells of vertebrates have their origin at the edges of the neural plate. They undergo an epithelial-to-mesenchymal transition which allows them to leave the midline, migrating away from it on either side. Neural crest cells are guided to various sites by interactions with the extracellular matrix over which they are moving, as well as by cell–cell interactions. Neural crest cells give rise to a wide variety of different cell types and these include nerve cells, cartilage of the face, and pigment cells.

Directed dilation is an important force in plants, and results from an increase in hydrostatic pressure inside a cell. Cell enlargement is a major process in plant growth and morphogenesis, providing up to a fiftyfold increase in the volume of a tissue. The driving force for expansion is the hydrostatic pressure exerted on the cell wall as a result of the entry of water into cell vacuoles by osmosis. Plant-cell expansion involves synthesis and deposition of new cell-wall material, and is an example of directed dilation. The direction of cell growth is determined by the orientation of the cellulose fibrils in the cell wall.

Chapter 5
Germ cells and sex

Animal embryos develop from a single cell, the fertilized egg or zygote, which is the product of the fusion of an egg and a sperm. In sexually reproducing organisms, there is a fundamental distinction between the germ cells, and the somatic body cells. The former give rise to the eggs and sperm and so determine the nature of the next generation, whereas body cells make no genetic contribution to the next generation. Germ cells have three key functions: the preservation of the genetic integrity of the germline; the generation of genetic diversity; and the transmission of genetic information to the next generation. In all but the simplest animals, the cells of the germline are the only cells that can give rise to a new organism. So, unlike body cells, which eventually all die, germ cells in a sense outlive the bodies that produced them. They are, therefore, very special cells and are prevented from ageing.

The outcome of germ-cell development in animals is either sperm or an egg. The egg is a particularly remarkable cell, as it ultimately gives rise to all the cells in an organism. In species whose embryos receive no nutrition from the mother after fertilization, the egg must also provide everything necessary for development, as the sperm contributes virtually nothing to the organism other than its chromosomes with their genes.

In animals, germline cells are specified and set apart in the early embryo, although functional mature eggs and sperm are only produced in the adult. An important property of germ cells is that they remain pluripotent—able to give rise to all the different types of cells in the body. Nevertheless, eggs and sperm in mammals have certain genes differentially switched off during germ-cell development by a process known as genomic imprinting, as discussed later. It is worth noting that some simple animals, such as *Hydra* discussed later, can reproduce asexually, by budding, and that even in some vertebrates, such as turtles, the eggs can develop without being fertilized. Plants, although reproducing sexually, differ from most animals in that their germ cells are not specified early in embryonic development, but during the development of the flowers. A special feature of plants is that single cells taken from the adult can give rise to a whole plant.

Germ cells differentiate into eggs and sperm within specialized reproductive organs called gonads: the ovary in females and the testis in males. In flies, nematodes, fish, and frogs, molecules localized in specialized cytoplasm in the egg are involved in specifying the germ cells. The clearest example of this is in the fly, where there is a region of special cytoplasm at the posterior pole of the egg that specifies germ cells. There is no evidence for special regions of the egg that specify germ cells in the chick or in the mouse and other mammals. In many animals, the primordial germ cells develop at some distance from the gonads, and only later migrate to them, where they differentiate into eggs or sperm. The earliest detectable primordial germ cells can be identified in the mouse just before the beginning of gastrulation and they form a cluster of six to eight cells. After about one week there are around 40 such cells in the primitive streak and these represent the full complement of primordial germ cells that will eventually migrate to the mouse gonads.

In order that the number of chromosomes is kept constant from generation to generation, germ cells are produced by a specialized type of cell division, called meiosis, which halves the chromosome number. Unless this reduction by meiosis occurred, the number of chromosomes would double each time the egg was fertilized. Germ cells thus contain a single copy of each chromosome and are called haploid, whereas germ-cell precursor cells and the other somatic cells of the body contain two copies and are called diploid. The halving of chromosome number at meiosis means that when egg and sperm come together at fertilization, the diploid number of chromosomes is restored.

Meiosis comprises two cell divisions; the chromosomes are replicated before the first division but not before the second, so that their number is halved. During an early stage of the first meiotic division, homologous chromosomes pair up and exchange regions, and this generates chromosomes with new combinations of genes. Meiosis thus results in gametes whose chromosomes carry different combinations of genes compared with the parent. This means that when sperm and egg come together at fertilization, the resulting animal will differ in genetic constitution from either of its parents. So, although we might resemble our parents, we never look exactly like them. A major error in human egg meiosis results in it having an extra chromosome 21, known as a trisomy of chromosome 21, as the result of an error at the first meiotic division. This trisomy is the cause of Down syndrome, and is one of the most common genetic causes of congenital malformations and learning disability.

The developing egg may rely on the synthetic activities of other cells. For example, the yolk proteins in birds and amphibians are made by liver cells, and are carried by the blood to the ovary, where they enter the developing egg (the oocyte) and become packaged into yolk platelets. Eggs vary enormously in size among different animals, but they are always larger than the body cells. In mammals, germ cells undergo a small number of cell divisions as they migrate to the gonad and never proliferate again after entry into meiosis; thus, the number of oocytes at this embryonic stage is generally considered to be the maximum number of eggs a female mammal can ever have. In humans, most oocytes degenerate before puberty, leaving about 400,000 out of an original 6–7 million to last a lifetime. This number declines with age, with the decline becoming steeper after the mid-30s until menopause, typically in the 50s. In mammals and many other vertebrates, oocyte development is held in suspension in the first stage of meiosis, but after birth and when the female becomes sexually mature, the oocytes start to undergo maturation, as the result of hormonal stimuli.

The development of sperm is quite different from that of eggs. Diploid germ cells that give rise to sperm do not enter meiosis in the embryo, but become arrested at an early stage of the cell cycle in the embryonic testis. They resume proliferation after birth. Later, in the sexually mature animal, the stem cells undergo meiosis and develop into sperm. Thus, unlike the fixed number of eggs in female mammals, sperm continue to be produced throughout the life of the organism.

Certain genes in eggs and sperm are **imprinted**, so that the activity of the same gene is different depending on whether it is of maternal or paternal origin. Improper imprinting can lead to developmental abnormalities in humans. At least 80 imprinted genes have been identified in mammals, and some are involved in growth control. For example, the insulin-like growth factor IGF-2 is required for embryonic growth; in the maternal genome, its gene is turned off (imprinted), so that only the paternal copy of the gene is active. A father benefits from maximal growth for his own offspring so that his genes have a good chance of surviving and being carried on, while the mother, who may mate with different males, benefits more by spreading her resources over all her offspring, and so needs to prevent too much growth in any one embryo. Thus, a gene like that for IGF-2 that promotes embryonic growth is turned off in the mother's copy. There are, however, many effects of imprinted genes other than on growth.

A number of developmental disorders in humans are associated with imprinted genes. Infants with Prader-Willi syndrome fail to thrive and later can become extremely obese; they also show mental retardation and mental disturbances such as obsessional-compulsive behaviour. Angelman syndrome results in severe motor and mental retardation. Beckwith-Wiedemann syndrome is due to a generalized disruption of imprinting on a region of chromosome 7 and leads to excessive foetal overgrowth and an increased predisposition to cancer.

Fertilization is the fusion of egg and sperm and is the trigger that initiates development. Fertilization and egg activation are associated with an explosive release of free calcium ions that initiates the completion of meiosis in the fertilized egg by acting on proteins that control the cell division. The nuclei of the egg and sperm then fuse to form the nucleus of the embryo and the egg starts dividing and embarks on its developmental program.

Sperm are motile cells, typically designed for activating the egg and delivering their nucleus into the egg cytoplasm. They essentially consist of a nucleus, mitochondria to provide an energy source, and a flagellum for movement. The sperm contributes virtually nothing to the organism other than its chromosomes. In mammals, sperm mitochondria are destroyed following fertilization, and so all mitochondria in the animal are of maternal origin.

In many marine organisms, such as sea urchins, sperm released into the water by the males are attracted to eggs by a gradient of chemical released by the egg. The membranes of the egg and sperm fuse, and the sperm nucleus enters the egg cytoplasm. In mammals and many other animals, of all the sperm released by the male, only one fertilizes each egg. In many animals, including mammals, sperm penetration activates a blocking mechanism in the egg that prevents any further sperm entering. This is necessary because if more than one sperm nucleus enters the egg there will be additional sets of chromosomes, resulting in abnormal development. In humans, embryos with such abnormalities fail to develop. Specializations of the egg are directed to preventing fertilization by more than one sperm, and the unfertilized egg is usually surrounded by several protective layers outside of the cell membrane. Different organisms have different ways of ensuring fertilization by only one sperm. In birds, for example, many sperm penetrate the egg but only one sperm nucleus fuses with the egg nucleus; the other sperm nuclei are destroyed in the cytoplasm.

In mammals there are very few mature eggs—usually one or two in humans and about 10 in mice—waiting to be fertilized, and fewer than a hundred of the millions of sperms deposited actually reach these eggs. Human and other mammalian eggs can be fertilized in culture and the very early embryo transferred to the mother's womb, where it implants and develops normally. This procedure of *in vitro* fertilization (IVF) has been of great help to couples who have, for a variety of reasons, difficulty in conceiving. We now take IVF as a treatment for human infertility almost for granted, even though it is only 30 years since the first IVF baby, Louise Brown, was born in the United Kingdom. A human egg can even be fertilized by injecting a single intact sperm directly into the egg in culture, which is useful when infertility is due to the sperm being unable to penetrate the egg. IVF embryos can be frozen and then successfully implanted many years later.

More recently, it has become possible to screen the genes of embryos produced by IVF, with the aim of avoiding the implantation of an embryo carrying a hereditary genetic defect. Because of the capacity of human embryos to regulate, one cell can be removed from an embryo during early cleavage without affecting its subsequent development. The DNA from this single cell is then screened for mutations known to cause disease. This preimplantation screening has been most widely carried out in cases where parents are known carriers for a particular genetic disease, such as cystic fibrosis. Such screening ensures that only those embryos with the normal gene are implanted into the mother. The demand for preimplantation diagnosis is increasing because it can be used to screen not only for mutations in genes that affect a newborn baby but also in genes that predispose an individual to disease in later life. An example is *BRCA1*, mutations in which predispose women to develop breast and ovarian cancer and account for 80% of these tumours in women with a genetically inherited predisposition. In males, mutations in *BRCA1* are linked to an increased susceptibility to prostate cancer. By screening embryos for mutations in *BRCA1*, the genetic predisposition to these cancers can be eliminated from a family. Preimplantation genetic diagnosis raises some practical and ethical questions, such as which genetic diseases should be screened for. Even when screening is not at issue, IVF usually produces more embryos than are implanted. What should happen to these spare embryos is a matter of debate, but most are disposed of.

Early development is similar in both male and female mammalian embryos, with sexual differences only appearing at later stages. The development of the individual as either male or female is genetically fixed at fertilization by the chromosomal content of the egg and sperm that fuse to form the fertilized egg. There are two sex chromosomes, X and Y. Females have cells with two X chromosomes (XX) while males have an X and a Y (XY). Each sperm carries either an X or Y chromosome, while the egg has an X. The genetic sex of a mammal is thus established at the moment of conception, when the sperm introduces either an X or a Y chromosome into the egg. A gene on the Y chromosome, *SRY*, causes testes to develop, which secrete hormones like testosterone that cause the development of male tissues and suppress female development. The development of a penis and a scrotum in males instead of the clitoris and labia of females and the reduced size of mammary glands in males are due to the action of the hormone testosterone (Figure 19). In other animals, like the fly, the number

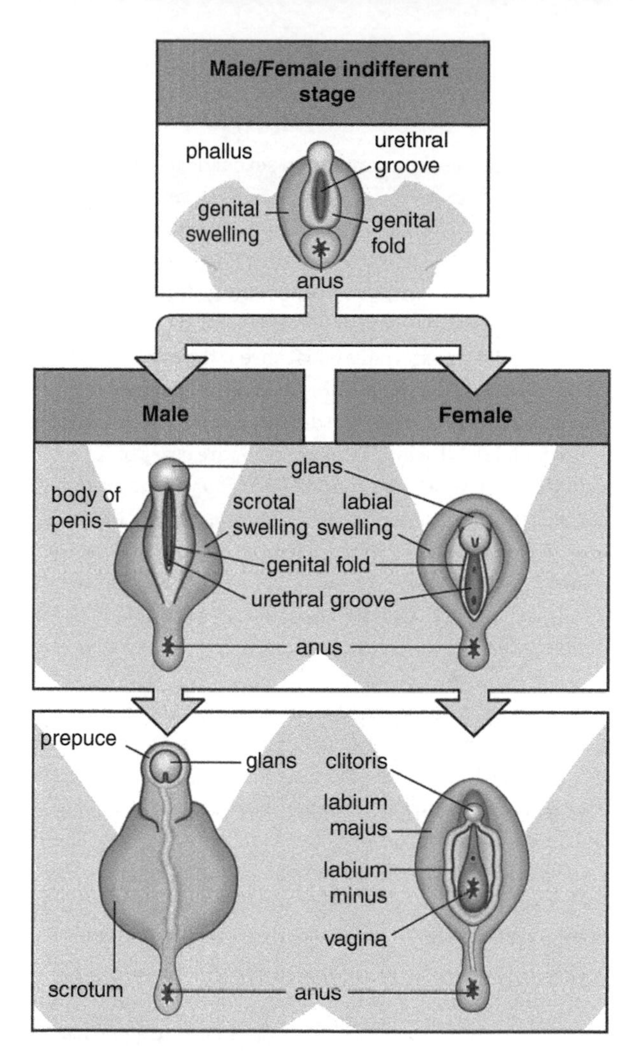

19. Development of the genitalia in humans. At an early embryonic stage, the genitalia are the same in males and females. After testis formation in males, the phallus and the genital fold give rise to the penis, whereas in females they give rise to the clitoris and the labia minus. The genital swelling forms the scrotum in males and the labia majus in females

of XX chromosomes in each cell determines their sex and hormones are not involved.

The role of hormones in mammalian sexual development is illustrated by rare cases of abnormal sexual development. Certain XY males develop as females in external appearance, even though they have testes and secrete testosterone if they have a mutation that renders them insensitive to testosterone. Conversely, genetic females with a completely normal XX constitution can develop as phenotypic males in external appearance if they are exposed to male hormones during their embryonic development. In the absence of a Y chromosome, the default development of tissues is along the female pathway. There are also rare cases of XY individuals who are female, and XX individuals who are physically male. This is due to part of the Y chromosome being lost in XY females or to part of the Y chromosome being transferred to the X chromosome in XX males. This can happen during meiosis in the male germ cells as the X and Y chromosomes are able to pair up, and exchange can occur between them.

In many animals, like mammals including us, there is an imbalance of X-linked genes between the sexes. One sex has two X chromosomes, whereas the other has just one. This imbalance has to be corrected to ensure that the level of expression of genes carried on the X chromosome is the same in both sexes. The mechanism by which the imbalance in X-linked genes is dealt with is known as dosage compensation. Failure to correct the imbalance leads to abnormalities and arrested development. Mammals, such as mice and humans, achieve dosage compensation in females by inactivating one X chromosome, selected randomly, in each cell. Once an X chromosome has been inactivated in an embryonic cell, this chromosome is maintained in the inactive state in all the body cells, and inactivation persists throughout the life of the organism. The mosaic effect of X-inactivation is sometimes visible in the coats of female mammals. Female mice in which one pigment gene on the X chromosome is inactive have patches of colour on their coat, produced by clones of epidermal cells that express the X chromosome carrying a functional pigment gene. Dosage compensation in the fly works in a different way. Instead of repression of the 'extra' X activity in females, transcription of the X chromosome in males is increased nearly twofold. In the nematode, dosage compensation is achieved by reducing the level of the X chromosome expression in XX individuals to that of the single X chromosome in males.

Unlike animals, plants do not set aside germ cells in the embryo and germ cells are only specified when a flower develops. Any meristem cell can, in principle, give rise to a germ cell of either sex, and there are no sex chromosomes. The great majority of flowering plants give rise to flowers that contain both male and female sexual organs, in which meiosis occurs. The male sexual organs are the stamens; these produce pollen, which contains the male gamete nuclei corresponding to the sperm of animals. At the centre of the flower are the female sex organs, which consist of an ovary of two carpels, which contain the ovules. Each ovule contains an egg cell. When a pollen grain is deposited on the carpel surface, it grows a tube that penetrates the carpel and delivers two haploid pollen nuclei to an ovule. One nucleus fertilizes the egg cell while the other fuses with two other nuclei in the ovule. This forms a triploid cell that will develop into a specialized nutritive tissue—the endosperm—that surrounds the fertilized egg cells and provides the food source for embryonic development.

Chapter 6
Cell differentiation and stem cells

The development of many different cell types such as muscle, blood, and skin is known as cell differentiation. It occurs first in the developing embryo and continues after birth and throughout adulthood. The character of specialized cells such as nerve, muscle, or skin is the result of a particular pattern of gene activity that determines which proteins are synthesized. There are more than 200 clearly recognizable differentiated cell types in mammals. How these particular patterns of gene activity develop is a central question in cell differentiation. Gene expression is under a complex set of controls that include the actions of transcription factors, and chemical modification of DNA. External signals play a key role in differentiation by triggering intracellular signalling pathways that affect gene expression.

Embryonic cells start out looking similar to each other but become different, acquiring distinct identities and specialized functions. Early embryonic cells fated to become different cell types differ primarily from each other only in their pattern of gene activity, and thus in the proteins present. Differentiation occurs over successive cell generations, the cells gradually acquiring new features while their potential fates become more and more restricted. The early precursors of cartilage and muscle cells have no obvious structural differences from each other and so look the same and might be described as undifferentiated; nevertheless, they will differentiate as cartilage and muscle, respectively, when cultured under appropriate conditions. Similarly, at an early stage in their differentiation, the precursors of white blood cells are structurally indistinguishable from those of red blood cells, but are distinct in the proteins they express.

As with earlier processes in development, the central feature of cell differentiation is a change in gene expression, which brings about a change in the proteins in the cells. The genes expressed in a differentiated cell include not only those for a wide range of 'housekeeping' proteins, such as the enzymes involved in energy metabolism, but also genes encoding cell-specific proteins that characterize a fully differentiated cell: hemoglobin in red blood cells, keratin in skin epidermal cells, and muscle-specific actin and myosin protein filaments in muscle. Expression of a single protein can change a cell's differentiated state. If the gene *myoD* is introduced into fibroblasts—connective tissue cells—for example, they will develop into muscle cells, as *myoD* encodes a master transcriptional regulator of muscle differentiation. It is nevertheless important to realize that several thousand different genes are active in any given cell in the embryo at any one time, though only a small number of these may be involved in specifying cell fate or differentiation. Special techniques can detect all the genes that are being expressed in a particular tissue or at a particular stage of development.

Cell differentiation is known to be controlled by a wide range of external signals but it is important to remember that, while these external signals are often referred to as being 'instructive', they are 'selective', in the sense that the number of developmental options open to a cell at any given time is limited. These options are set by the cell's internal state which, in turn, reflects its developmental history. External signals cannot, for example, convert an endodermal cell into a muscle or nerve cell. Most of the molecules that act as developmentally important signals between cells during development are proteins or peptides, and their effect is usually to induce a change in gene expression. These proteins and peptides bind to receptors in the cell membrane and the signal is relayed to the cell nucleus by intracellular signalling pathways—signal transduction. The same external signals can be used again and again with different effects because the cells' histories are different.

Each cell in the body of a multicellular organism contains a nucleus derived from the single nucleus in the fertilized egg. But patterns of gene activity in differentiated cells vary enormously from one cell type to another. To understand the molecular basis of cell differentiation, we first need to know how a gene can be expressed in a cell-specific manner. Why does a certain gene get switched on in one cell and not in another? Most of the key genes controlling development are initially in an inactive state and require activating transcription factors to turn them on. These activators bind to specific regulatory control regions in the DNA, which are often called enhancers. For any given gene, the specificity of its activation is due to particular combinations of gene-regulatory proteins binding to individual sites in the control regions (see Figure 4). At least 1,000 different transcription factors are encoded in the genomes of the fly and the nematode, and as many as 3,000 in the human genome. On average, around five different transcription factors act together at a control region and, in some cases, considerably more. As well as sites that bind activators, control regions may contain sites that bind repressors, proteins that inhibit gene expression; these prevent a gene being expressed at the wrong time or in the wrong place. In general, it can be assumed that activation of each gene involves a unique combination of transcription factors. In vertebrates, chemical modification at certain sites in the DNA is correlated with the blocking of transcription in those regions and this provides a mechanism for passing on a blocked pattern of gene activity to daughter cells. This is also known as epigenetics, and can even persist over into the next generation when the genes of germ cells are involved.

Stem cells involve some special features in relation to differentiation. A single stem cell can divide to produce two daughter cells, one of which remains a stem cell while the other gives rise to a lineage of differentiating cells. This occurs in our skin and gut all the time and also in the production of blood cells. It also occurs in the embryo. One basis for this behaviour is that there is an intrinsic difference between the two daughter cells because the stem-cell division is asymmetric, resulting in the two cells acquiring a different complement of proteins. The second possibility is that external signals make the daughter cells different: a daughter cell that remains in a stem-cell niche continues to renew itself because of signals from the local cells, whereas one that ends up outside the niche differentiates. Embryonic stem (ES) cells from the inner cell mass of the early mammalian embryo when the primitive streak forms, can, in culture, differentiate into a wide variety of cell types, and have potential uses in regenerative medicine. As will be discussed, it is now possible to make adult body cells into stem cells, which has important implications for regenerative medicine.

Hematopoiesis, or blood formation, is a particularly well-studied example of cell differentiation. Hematopoietic stem cells are multipotent—they can give rise to a range of differentiated cell types. The existence of the multipotent stem cell for blood formation can be inferred from the ability of cells in the bone marrow to reconstitute a complete blood and immune system when transplanted into individuals whose own bone marrow has been destroyed, a property that is exploited therapeutically in the use of bone marrow transplantation to treat diseases of the blood and immune system. The bone marrow contains multipotent stem cells that become irreversibly committed to one or other of the lineages leading to the different blood cell types. All this activity occurs in the microenvironment of the bone marrow and is regulated by external signals. Hematopoiesis is, in effect, a complete developmental system in miniature, in which a single multipotent stem cell gives rise to numerous different blood cell types. There is a continual turnover of blood cells, so that hematopoiesis must continue throughout life. As a measure of the complexity of hematopoiesis, the cells involved have been found to express at least 200 transcription factors, a similar number of membrane-associated proteins, and around 150 signalling molecules.

A main feature of red-blood-cell differentiation is the synthesis of large amounts of the oxygen-carrying protein hemoglobin, which involves the coordinated regulation of two different sets of globin genes by transcription factors. All the hemoglobin contained in a fully differentiated red blood cell is produced before its terminal differentiation when, in mammals, the nucleus is expelled from the cell. Vertebrate hemoglobin is made up of two identical α-type and two identical β-type globin chains. In mammals, different members of each family are expressed at various stages of development so that distinct hemoglobins are produced during embryonic, foetal, and adult life. This is a mammalian adaptation to differing requirements for oxygen transport at different stages of life; human foetal hemoglobin, for example, has a higher affinity for oxygen than adult hemoglobin, and so is able to efficiently take up oxygen for the embryo. This developmentally regulated expression of hemoglobin depends on a region a very long distance upstream from the gene that loops in such a way that proteins binding to it can contact and interact with proteins bound to the hemoglobin gene promoters. Mutations in the globin genes are the cause of several relatively common inherited blood diseases. One of these, sickle-cell anaemia, is caused by a point mutation (a single nucleotide change in the gene). In people with both copies of this mutation, the abnormal hemoglobin molecules aggregate into fibres, forcing the cells into the characteristic sickle shape. These blood cells cannot pass easily through fine blood vessels and tend to block them, causing many of the symptoms of the disease. They also have a much shorter lifetime than normal blood cells. Together these effects cause anaemia—a lack of functional red blood cells. Sickle-cell anaemia is one of the few genetic diseases where the link between a mutation and the subsequent developmental effects on health is fully understood. Individuals with just one copy mutated are more resistant to malaria.

Other cell types that come from that blood stem cell include macrophages, which are cells that clean up the debris around cells, particularly those that have died, and the white blood cells (lymphocytes) of the immune system that give rise to antibodies.

Mammalian skin is composed of two cellular layers: the dermis, which mainly contains connective tissue cells known as fibroblasts; and the protective outer epidermis, which contains mainly keratin-filled cells. A basement membrane separates the two cell layers. Because of the protective function of the epidermis, cells are continuously lost from its outer surface and must be replaced—every four weeks we have a brand new epidermis—and it is maintained throughout life by stem cells that reside in the basal layer of the epidermis (Figure 20). Once a cell leaves this stem-cell layer, it becomes committed to differentiate, and asymmetrical divisions of basal cells lead to one daughter remaining in the basal layer and the other becoming committed to being a keratin-containing cell. The dead cells eventually flake off the surface. The epithelial cells lining our gut are also continuously replaced from stem cells.

Differentiation of vertebrate skeletal muscle can be studied in cell culture and provides a valuable model system. Skeletal muscle cells derive from a region of the somites. Myoblasts, cells that are committed to forming muscle, can be isolated from chick or mouse somites to produce cell cultures where the cells will proliferate until growth factors are removed and differentiation into muscle cells begins. The cells then begin to synthesize muscle-specific proteins and also undergo structural changes. They first become bipolar in shape, and then fuse to form large, tube-like cells with multiple nuclei that develop into muscle. Once

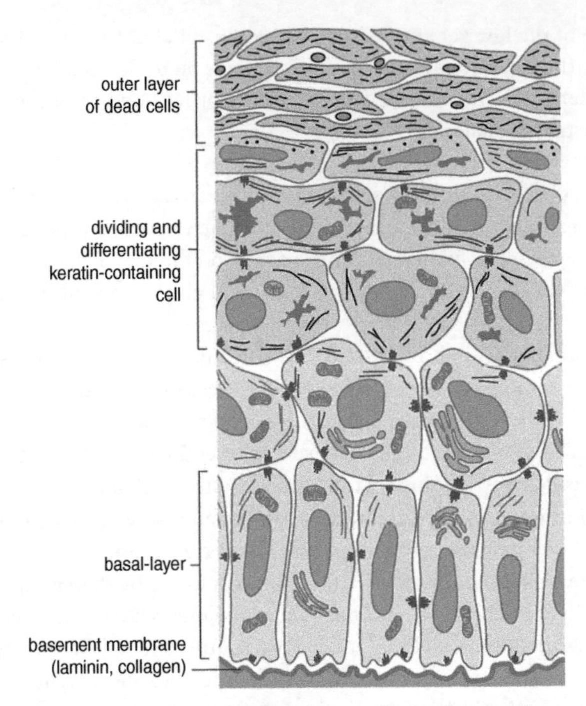

20. Differentiation of skin cells from stem cells in the basal layer

formed, skeletal muscle cells can enlarge by cell growth but do not divide. In adult mammalian muscle there are satellite stem cells that can divide and differentiate into new muscle cells if the muscle is damaged.

Cells can commit suicide during development. Programmed cell death—**apoptosis**—is a biological process quite different from the cell death that occurs if the cell is damaged. While not strictly cell differentiation, it is convenient to consider it here in these terms. It is involved, for example, in the development of the vertebrate limb, where the death of cells between the developing digits is essential for separating the digits. The development of the vertebrate nervous system also involves the death of large numbers of neurons. Programmed cell death is particularly important in the development of the nematode: 959 somatic cells come from the egg and 131 die during development. In all these cases, the dying cell undergoes a type of cell suicide that requires both RNA and protein synthesis. The dying cell breaks up into fragments, which are finally removed by macrophages. These features distinguish apoptosis from cell death due to damage, where the whole cell tends to swell and eventually burst open. The cells in all tissues are intrinsically programmed to undergo cell death, and are only prevented from dying by positive control signals from neighbouring cells. Programmed cell death also plays a key role both in controlling growth and in preventing the development of cancer cells.

How reversible is cell differentiation? To what extent can the pattern of gene activity in differentiated cells revert to that found in the fertilized egg? One way of finding out whether it can be reversed is to place the nucleus of a differentiated cell in a different cytoplasmic environment, one that contains a different set of gene-regulatory proteins. This experiment led to cloning. The most dramatic experiments addressing the reversibility of differentiation have investigated the ability of nuclei from cells at different stages of development to replace the nucleus of an egg and to support normal development. If they can do this, it would indicate that no irreversible changes have occurred to the genome during differentiation. It would also show that a particular pattern of nuclear gene activity is determined by whatever transcription factors and other regulatory proteins are being synthesized in the cytoplasm of the cell.

Such experiments were first carried out using frog eggs, which are particularly robust for experimental manipulation. In the unfertilized frog egg, the nucleus lies directly below the surface at the animal pole. A dose of ultraviolet radiation directed at the animal pole destroys the DNA within the nucleus, and thus prevents all genes functioning. These effectively enucleated eggs can then be injected with a nucleus taken from a cell at a later stage of development, or even an adult cell, to see whether it can function in place of the inactivated nucleus. The results are striking: nuclei from early embryos and from some types of cells of adults such as gut and skin epithelial cells, can replace the egg nucleus and support the development of an embryo up to the tadpole stage and, in a small number of cases, even into an adult. The process by which they are produced is called cloning as it makes an animal with the identical set of genes to that of the cell that donated the nucleus. However, the success rate with nuclei from the body cells of adults is very low, with only a small percentage of nuclear transplants developing past the cleavage stage. These results show that the genes required for development are not irreversibly altered during development and the behaviour of cells is determined entirely by factors present in the cell.

What about organisms other than the frog? The first mammal to be cloned was a lamb, the famous Dolly. In this case, the nucleus was taken from a cell line derived from the udder. In general, the success rate of cloning by body-cell nuclear transfer in mammals is low, and the reasons for this are not yet well understood. However a variety of mammals, including cattle, sheep, dogs, and a camel, have been cloned, although no primate like a monkey has yet been cloned and developed into an adult. Most cloned mammals derived from nuclear transplantation are usually abnormal in some way. The cause of failure is incomplete reprogramming of the donor nucleus to remove all the earlier modifications. A related cause of abnormality may be that the reprogrammed genes have not gone through the normal imprinting process that occurs during germ-cell development, where different genes are silenced in the male and female parents. The abnormalities in adults that do develop from cloned embryos include early death, limb deformities and hypertension in cattle, and immune impairment in mice. All these defects are thought to be due to abnormalities of gene expression that arise from the cloning process. Studies have shown that some 5% of the genes in cloned mice are not correctly expressed and that almost half of the imprinted genes are incorrectly expressed. It is because the child will almost certainly be abnormal that cloning of humans should be prevented, and not because of ethical issues, of which there are none. Despite reports in the media that humans have been cloned, none of these reports has, fortunately, been verified.

We have already seen several examples of multipotent stem cells that both self-renew and give rise to a range of different cell types. If such stem cells could be produced reliably and in sufficient numbers, it might be possible to use them to replace cells that have been damaged or lost by disease or injury. This is one of the main aims of the field of regenerative medicine. The therapeutic use of stem cells will depend on understanding precisely how gene activity can be controlled in stem cells to give the desired cell type, and just how plastic stem cells are.

The primary examples of pluripotent stem cells in mammals are the embryonic stem cells (ES cells) derived from the inner cell mass of the early embryo. Mouse ES cells have been studied intensively. They can be maintained in culture for long periods, apparently indefinitely, but if injected into an early embryo that is then returned to the mother, they can contribute to all the types of cells in that embryo (Figure 21). To maintain mouse ES cells in a pluripotent state in cell culture, the cells must express a particular combination of four transcription factors whose expression together is restricted to pluripotent stem cells. ES cells can be made to differentiate into a particular cell type by manipulating the culture conditions, particularly in respect of the growth factors present. Under specific treatments, ES cells can differentiate into heart muscle, blood cells, neurons, pigmented cells, epithelia, fat cells, macrophages, and even germ cells.

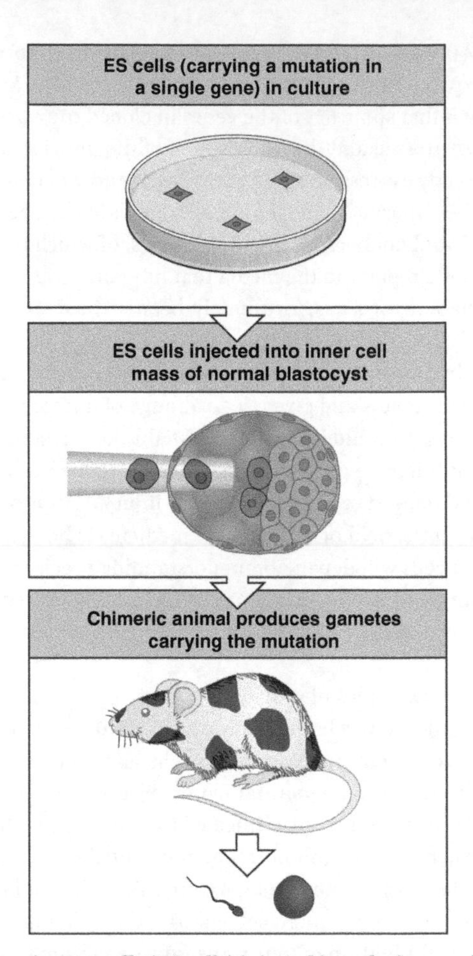

21. Embryonic stem cells (ES cells) injected into the inner cell mass of a blastocyst can give rise to all cell types

The goal of regenerative medicine is to restore the structure and function of damaged or diseased tissues. As stem cells can proliferate and differentiate into a wide range of cell types, they are strong candidates for use in cell-replacement therapy, the restoration of tissue function by the introduction of new healthy cells. This type of therapy might eventually offer an alternative to conventional organ transplantation from a donor, with its attendant problems of rejection and shortage of organs, and might also be able to restore function to tissues such as brain and nerves. There are claimed to be ethical issues associated with the use of human ES cells, as the embryo can be destroyed when the stem cells are taken from it, and there are those who believe that this is a destruction of a human life. There is good evidence that the embryo does not necessarily represent an individual at this very early stage, as it is still capable of giving rise to twins at a later stage. And in practice, many early embryos are lost during assisted reproduction involving IVF, a widely accepted medical intervention. Acceptance of IVF and rejection of the use of ES cells could be seen as a contradiction.

The generation of insulin-producing pancreatic β cells from ES cells to replace those destroyed in type 1 diabetes is a prime medical target. Treatments that direct the differentiation of ES cells towards making endoderm derivatives such as pancreatic cells have been particularly difficult to find. Nevertheless, using knowledge of the signals that induce endoderm and pancreas development in mouse embryos, progress has been made in devising methods for differentiating human ES cells into pancreatic progenitor cells. Similar strategies can also be used in other diseases. The neurodegenerative Parkinson disease is another medical target.

The use of pluripotent stem cells taken from a patient would avoid the ethical issues associated with embryonic stem cells, as well as avoiding the problem of immune rejection of transplanted cells.

To generate such stem cells of the patient's own tissue type would be a great advantage, and the recent development of induced pluripotent stem cells (iPS cells) offers the most exciting new opportunities. iPS cells were made from fibroblasts by introducing and expressing genes for the four transcription factors that are associated with pluripotency in ES cells. There is the risk of tumour induction in patients undergoing cell-replacement therapy with ES cells or iPS cells; undifferentiated pluripotent cells introduced into the patient could cause tumours. Only stringent selection procedures that ensure no undifferentiated cells are present in the transplanted cell population will overcome this problem. And it is not yet clear how stable differentiated ES cells and iPS cells will be in the long term.

Chapter 7
Organs

Once the basic animal body plan has been laid down, the development of organs as varied as limbs and eyes can begin. Organ development involves large numbers of genes and, because of this complexity, general principles can be quite difficult to distinguish. Nevertheless, many of the mechanisms used in organogenesis are similar to those of earlier development, and certain signals are used again and again. Pattern formation in development in a variety of organs can be specified by **positional information**, which is specified by a gradient in some property. The basic idea is illustrated by the French flag model: if the cells know their positions, then they can interpret them by differentiating into red, white, or blue, so as to make the flag. This model has the advantage that the pattern will be the same for different sizes. Cells might know their position by the concentration of a graded molecule across the field.

A specialized group of cells at one boundary of the area to be patterned could send a signal, possibly the concentration of a molecule, which decreases with distance from the source, thus forming a gradient of information. Any cell along the gradient then 'reads' the concentration at that particular point, and interprets this to respond in a manner appropriate to its position by switching on a particular pattern of gene expression. Molecules that have such a graded concentration and which can induce such changes in cell fate are known as **morphogens**. Gradients of positional information have been proposed for various patterns during development: regionalization of the antero-posterior and dorso-ventral axes and patterning of segments and imaginal discs in insects; mesoderm patterning in vertebrates; vertebrate limb patterning; and patterning along the dorso-ventral axis of the vertebrate neural tube, among others. It is still unclear how gradients of positional information are formed, particularly the relative roles of morphogen diffusion and cell–cell interactions.

The vertebrate embryonic limb is a particularly good system in which to study development of an organ, as the basic pattern is initially quite simple. The basic principles of limb patterning have been most extensively investigated in chick embryos, because the developing limbs themselves are easily accessible for microsurgical manipulation. Mice are used to study some aspects of limb development, mainly through spontaneous and artificial mutants. In chick embryos, the first signs of wings are small protrusions—the limb buds—that arise from the body wall of the embryo. The elements that make up the limb skeleton are first formed as cartilage and are later replaced by bone, and muscles and tendons will also develop. The limb has three developmental axes: the proximo-distal axis runs from the shoulder of the limb to the tip; the antero-posterior axis runs across the digits from the thumb to the little finger, in the chick wing that is from digit 2 to digit 4; and the dorso-ventral axis runs from the back of the hand to the palm. The early limb bud has a core of loose proliferating cells contained within an outer layer of ectodermal cells. The bones and tendons develop from these loose cells, but the limb muscles have a separate lineage—they migrate into the limb bud from the somites.

At the tip of the limb bud is a thickening in the ectoderm—the apical ridge—which causes a dorso-ventral flattening of the limb bud. Directly beneath the apical ectodermal ridge lies a region that is composed of rapidly proliferating undifferentiated cells

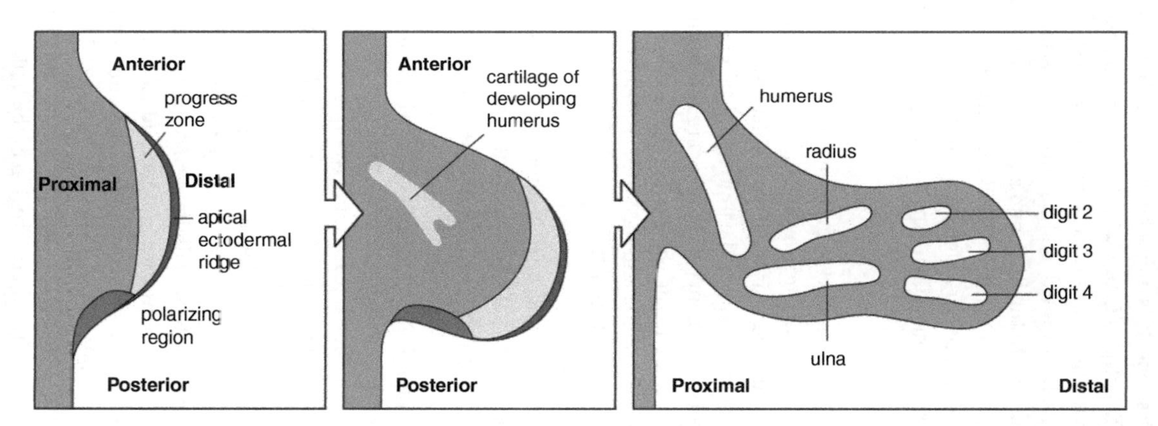

22. **There are two signalling regions in the chick wing bud: the polarizing region and the apical ridge. The progress zone is beneath the apical ridge. The limb develops in a proximo-distal direction**

called the progress zone. It is only when cells leave this zone that they begin to differentiate (Figure 22). The two main organizing regions of a limb are the apical ridge, which provides signals that are essential for limb outgrowth and for correct patterning along the proximo-distal axis of the limb, and a polarizing region, which is a group of cells on the posterior side of the limb bud that is crucial for determining pattern along the antero-posterior limb axis. Cells of the polarizing region express the signalling protein Sonic hedgehog.

As the bud grows, the cells start to differentiate and cartilaginous structures begin to appear. The proximal part of the limb—that is, the part nearest to the body—is the first to differentiate, and differentiation proceeds towards the limb tip (distally) as the limb extends. The cartilage elements of the wing are laid down in a proximo-distal sequence in the chick wing: the humerus, radius, and ulna; the wrist elements; and then the three easily distinguishable digits, 2, 3, and 4.

The developing chick limb bud can be thought of as being determined by the positional information the cells acquire, though this is a bit controversial. The cells have their positions specified with respect to the main limb axes. The specification of pattern along the antero-posterior axis of the wing bud is best seen in relation to the three digits. Patterning along this axis is what specifies the individual digits and gives them their identities. The organizing region for the antero-posterior axis is the polarizing region. When the polarizing region from an early chick wing bud is grafted to the anterior margin of another early chick wing bud, a wing with a mirror-image pattern develops: instead of the normal pattern of digits—2 3 4—the pattern is 4 3 2 2 3 4 (Figure 23). The pattern of muscles and tendons in the limb shows similar mirror-image changes. The additional digits come from the host limb bud and not from the graft, showing that the grafted polarizing region has altered the developmental fate of the host cells in the anterior region of the limb bud. One way that the polarizing region could specify position along the antero-posterior axis is by producing a morphogen, that is a diffusible molecule, that forms a posterior to anterior gradient. The concentration of morphogen could specify the character of the digits. Digit 4 would develop at a high concentration, digit 3 at a lower one, and digit 2 at an even lower one. When a small number of polarizing region cells are grafted anteriorly, only an additional digit 2 develops. The leg-polarizing region has a similar effect. The limb buds of a number of other vertebrates, including mouse, pig, ferret, turtles, and even human limb buds, have been shown to have a polarizing region. When the posterior margin of a limb bud from an embryo of these species is grafted to the anterior margin of a chick wing bud, additional chick wing digits are produced. This nicely illustrates that the effect of a signal depends on the responding cells. Separation of the digits, including our own, is due to the programmed death of the cells between these digits' cartilaginous elements. The webbed feet of ducks and other waterfowl are simply the result of less cell death between the digits.

There is no evidence for the morphogen leading to digit formation as distinct from specifying their character. A random mixture of limb cells without any polarizing cells, when placed in the limb's ectodermal jacket, can develop cartilage elements including digits. This implies the existence of some self-organizing mechanism in the limb bud that is capable of setting up a regular set of digit-like elements, but cannot give them distinct identities. Such a mechanism could, for example, be based on the principle of reaction–diffusion, which has been proposed to be responsible for repeated patterns, such as the stripes on angelfish. There are self-organizing chemical systems of diffusing molecules that spontaneously generate spatial patterns of concentration of some of their molecular components. The initial distribution of the molecules is uniform, but over time the system forms wave-like patterns. This reaction–diffusion system was discovered by Alan Turing. Such a mechanism could thus generate periodic patterns

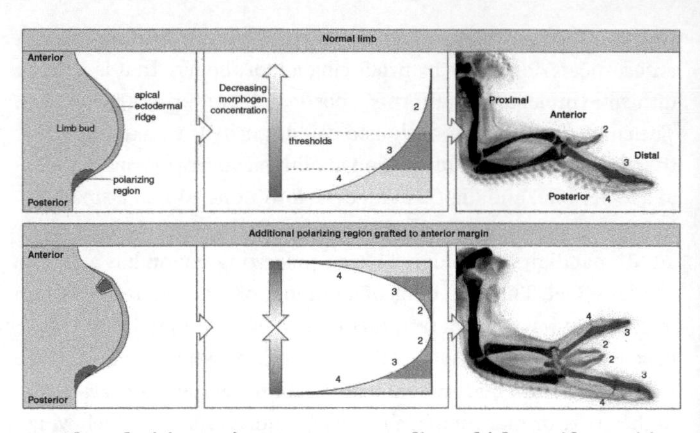

23. The polarizing region can set up a gradient which specifies position

such as the arrangement of digits, and could also generate pigment patterns in the skin of animals.

The mechanisms that pattern the proximo-distal axis are still a matter of some debate. Removing the apical ridge from a chick limb bud by microsurgery results in a significant reduction in outgrowth and the limb is truncated with distal parts missing. The earlier the ridge is removed, the greater the effect. A key signal from the ridge is provided by fibroblast growth factors. The longest-standing model proposes that proximo-distal patterning is specified by the length of time cells spend in the zone of cells at the tip of the limb beneath the apical ridge—the progress zone. As the limb bud grows, cells continually leave the zone. Because the limb bud extends from proximal to distal, the earliest cells leaving the zone develop into proximal elements and those leaving last form the tips of the digits. The model proposes that cells measure the time they spend in the progress zone and it is this that gives them their positional value along the proximo-distal axis. A timing mechanism of this sort for the limb is consistent with the observation that removal of the apical ridge results in a distally truncated limb, as the progress zone will no longer exist. Another line of evidence is that killing or blocking the proliferation of cells in the progress zone of a chick wing bud at an early stage, for example by irradiation with X-rays, results in the absence of proximal structures, whereas distal ones are present and can be almost normal. Because many cells in the irradiated progress zone do not divide, fewer cells than usual will leave the zone during early stages, so leading to the absence of proximal elements, but it gradually becomes repopulated with normal cells, and so distal structures do develop. Such a model could account for the absence of proximal limb structures and just a hand attached to the shoulder in babies born in the late 1950s and early 1960s to pregnant mothers who took the drug thalidomide to ease morning sickness. Thalidomide is known to interfere with the development of blood vessels, and this could have resulted in extensive cell death throughout the early limb bud including the progress zone. However, there are criticisms of these models and others have been proposed. The effects of both irradiation and thalidomide may, it has been suggested, be explained by their simply eliminating cartilage precursor cells at a time when proximal regions are forming but distal differentiation has not yet commenced.

The cells that give rise to limb muscles migrate into the limb bud from the somites and multiply, initially forming dorsal and ventral blocks of presumptive muscle. These blocks undergo a series of divisions to give rise to the individual muscles. Presumptive muscle cells do not acquire positional values in the same way as do cartilage and connective tissue cells, and all are equivalent. The muscle pattern is determined by the cells through which the future muscles migrate and this could be due to them binding to the migrating muscle cells.

The mechanism whereby the correct connections between tendons, muscles, and cartilage are established involves little or no specificity; they are truly democratic. If the tip of a developing wing is inverted dorso-ventrally, dorsal and ventral tendons can join up with inappropriate muscles and tendons; they simply make connections with those muscles and tendons nearest to their free ends—they are promiscuous.

The adult organs and appendages of the fly, such as the wings, legs, eyes, and antennae, develop from imaginal discs. The discs come from the embryonic ectoderm as simple pouches of epithelium during embryonic development and remain as such until metamorphosis when they develop into specific structures. Although all imaginal discs superficially look rather similar, they develop according to the segment in which they are located; their identity and their development is controlled by Hox genes. The wing and leg discs are initially specified as clusters of 20–40 cells and, during larval development, they grow about 1000-fold. The leg develops from a circular region in its imaginal disc (Figure 24). The centre of the disc becomes the tip of the leg as it extends upwards. At metamorphosis, by which time patterning of the imaginal discs is largely complete, wing and leg discs undergo a series of profound anatomical changes to form the adult structures.

The variety of colour markings on butterfly wings is remarkable: more than 17,000 species can be distinguished. Many of these patterns are variations on a basic 'ground plan' consisting of bands and concentric eyespots. Butterfly wings develop from imaginal discs in a similar way to fly wings. The eyespot is specified at a late stage in the development of the wing disc, and the pattern is dependent on a signal emanating from the centre of the spot. It is possible that eyespot development and distal leg patterning involve similar mechanisms and the eyespot may be thought of as a proximo-distal pattern superimposed on the two-dimensional wing surface. Different imaginal discs can have the same positional values, which implies that the wing and leg discs interpret positional signals in different ways. This interpretation is under the control of the Hox genes and can be illustrated in regard to the leg and antenna. If the Hox gene *Antennapedia*,

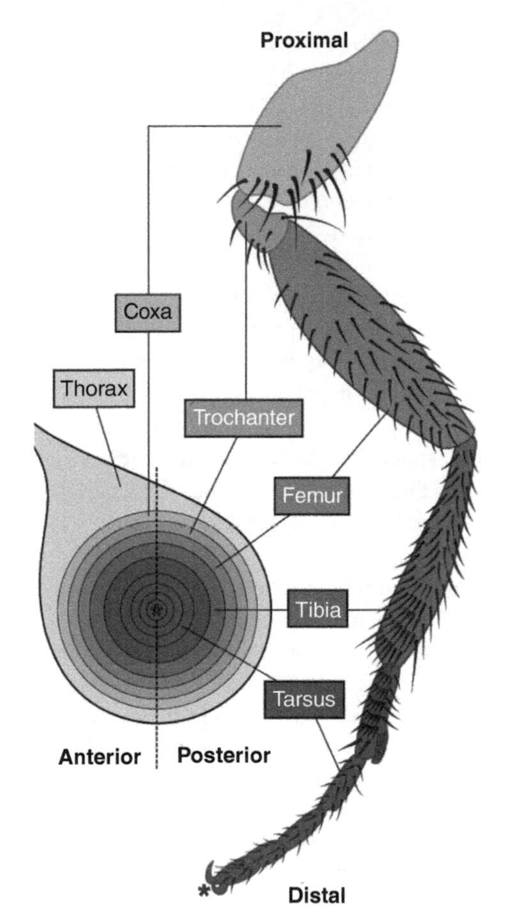

24. The fate map of the leg imaginal disc of the fruit fly, *Drosophila melanogaster*

which is normally expressed in parasegments 4 and 5 and specifies the discs for the second pair of legs, is expressed in the head region, the antennae develop as legs. It is possible to generate a clone of *Antennapedia*-expressing cells in a normal antennal disc. These cells develop as leg cells, but exactly which type of leg cell depends on their position along the proximo-distal axis; if, for example, they are at the tip, they form a claw. The positional values of the cells in the antenna and leg are thus the same, and the difference between the two structures lies in the interpretation of these values, which is governed by the expression or non-expression of the *Antennapedia* gene. Once evolution found a good way to make patterns, it used it again and again.

Structures as complex as the compound eyes of insects and the camera-like eyes of vertebrates are a remarkable achievement of evolution. They all contain a lens to focus the light, a retina composed of light-sensing photoreceptor cells, and a pigmented layer that absorbs stray light and prevents it interfering with photoreceptor signalling. Despite the great difference in the anatomy of the final eye, some of the same transcription factors specify eye formation in insects and vertebrates.

The vertebrate eye develops from the neural tube and the ectoderm of the head and is essentially an extension of the forebrain, with a contribution from the overlying ectoderm and migrating neural crest cells. The development of an eye starts with the formation of a bulge in the epithelial wall of the posterior forebrain to form an optic vesicle, which extends to meet the surface ectoderm. The optic vesicle interacts with the ectoderm to induce formation of the lens. After induction of the lens, the tip of the optic vesicle invaginates to form a two-layered cup, the inner epithelial layer of which will form the neural retina, while the outer layer will form the retinal pigment epithelium. The lens region then invaginates and detaches from the surface ectoderm to form a small hollow sphere of epithelium that will develop into the lens and the cells start to manufacture crystallin proteins. These cells eventually lose their internal structures to become completely transparent lens fibres filled with crystallin. The cornea is a transparent epithelium that seals the front of the eye.

A classic example of a gene with a conserved basic function is *Pax6*, which is required for the development of light-sensing structures in all animals with bilateral symmetry including the compound eyes of insects and the camera eyes of vertebrates. People with mutations in *Pax6* have a variety of eye malformations collectively known as aniridia. Amazingly, turning on expression of *Pax6* in a fly imaginal disc causes the development of fly-type compound eye.

The bodies of multicellular animals contain a multitude of tubes such as blood vessels, kidney tubules, and the branched airways in the mammalian lung. Many of these tubular systems undergo extensive branching during their development. The development of the fly tracheal system provides an excellent model for branching morphogenesis and has led to the identification of genes controlling the process that also act in the morphogenesis of the vertebrate lung. Air enters the tracheal system of the fly larva through small openings in the body wall, and oxygen is delivered to the tissues by some 10,000 or so fine tubules that develop during embryogenesis from 20 openings, 10 on each side. The ectoderm at the opening invaginates to form a hollow sac of around 80 cells which gives rise, through successive branching, to hundreds of fine terminal branches. Remarkably, the extension of the sacs to form branched tubes does not involve any further cell proliferation but is achieved by directed cell migration, cell rearrangement by intercalation, and changes in cell shape. As development proceeds, branches fuse to form a body-wide network of interconnected tubes. Unlike the fly tracheae, the outgrowth and branching of tubules in the vertebrate lung is the result of cell proliferation at the tip of the advancing tube, rather than cell migration. Nevertheless, the sprouting and outgrowth of tubules from the main tube depends, as in the fly, on the interaction of the tubular epithelium with signals from the surrounding mesodermal cells, and many of these are the same as in the fly.

Not surprisingly, the vascular system, including blood vessels and blood cells, is among the first organ systems to develop in vertebrate embryos, so that oxygen and nutrients can be delivered to the rapidly developing tissues. The defining cell type of the vascular system is the endothelial cell, which forms the lining of the entire circulatory system, including the heart, veins, and arteries. Blood vessels are formed by endothelial cells and these vessels are then covered by connective tissue and smooth muscle cells. Arteries and veins are defined by the direction of blood flow as well as by structural and functional differences; the cells are specified as arterial or venous before they form blood vessels but they can switch identity. The initial vessels are then elaborated into a vascular system ramifying throughout the body in which vessels extend and branch to form arteries, veins, and extensive networks of capillaries.

Differentiation of the vascular cells requires the growth factor VEGF (vascular endothelial growth factor) and its receptors, and VEGF stimulates their proliferation. Expression of the *Vegf* gene is induced by lack of oxygen and thus an active organ using up oxygen promotes its own vascularization. New blood capillaries are formed by sprouting from pre-existing blood vessels and proliferation of cells at the tip of the sprout. Cells at the tip extend filopodia-like processes that guide and extend the sprout. During their development, blood vessels navigate along specific paths towards their targets, with the filopodia at the leading edge responding to both attractant and repellent cues on other cells and in the extracellular matrix. Many solid tumours produce VEGF and other growth factors that stimulate vascular development and so promote the tumour's growth, and blocking new vessel formation is thus a means of reducing tumour growth.

The development of the vertebrate heart involves the specification of a mesodermal tube that is patterned along its long axis and is one of the first large organs to form in the embryo. It is first established as a single tube consisting of two epithelial layers one of which will give rise to heart muscle. During development, this tube becomes divided longitudinally into two chambers, the atrial and ventricular chambers. The later development of the heart involves asymmetric looping of the heart tube and is related to the left–right asymmetry of the embryo. A two-chambered heart is the basic adult form in fish, but in higher vertebrates, such as birds and mammals, looping and further partitioning give rise to the four-chambered heart. In humans, about 1 in 100 live-born infants has some congenital heart malformation, while *in utero*, heart malformation leading to death of the embryo occurs in between 5 and 10% of conceptions.

Flowers

Flowers contain the reproductive cells of higher plants and develop from the shoot meristem. In most plants, the transition from a shoot meristem producing leaves to a floral meristem that produces a flower is largely under environmental control, with day length and temperature being important determining factors. Flowers, with their arrangement of floral organs—sepals, petals, stamens, and carpels—are rather complex structures. The individual parts of a flower each develop from a floral organ primordium produced by the floral meristem. Unlike leaf primordia, which are all identical, the floral organ primordia must each be given a correct identity and be patterned according to it. An *Arabidopsis* flower has four concentric whorls of structures (Figure 25), which reflect the arrangement of the floral organ primordia in the meristem. The sepals (whorl 1) arise from the outermost ring of meristem tissue, and the petals (whorl 2) from a ring of tissue lying immediately inside it. An inner ring of tissue gives rise to the male reproductive organs—the stamens (whorl 3). The centre of the meristem develops into the female reproductive organs—the carpels (whorl 4). In a floral meristem of *Arabidopsis*, there are 16 separate primordia, giving rise to a flower with four sepals, four petals, six stamens, and a pistil made up of two carpels.

The primordia arise at specific positions within the meristem, where they develop into their characteristic structures. Like the homeotic selector genes that specify segment identity in the fly, mutations in floral identity genes cause homeotic mutations in which one type of flower part is replaced by another. In an *Arabidopsis* mutant, for example, the sepals are replaced by carpels, and the petals by stamens. Such mutations identified the floral organ identity genes, and have enabled their mode of action to be determined. These mutant forms can be accounted for by an elegant model in which overlapping patterns of gene activity specify floral organ identity in a manner highly reminiscent of the way in which fly homeotic genes specify

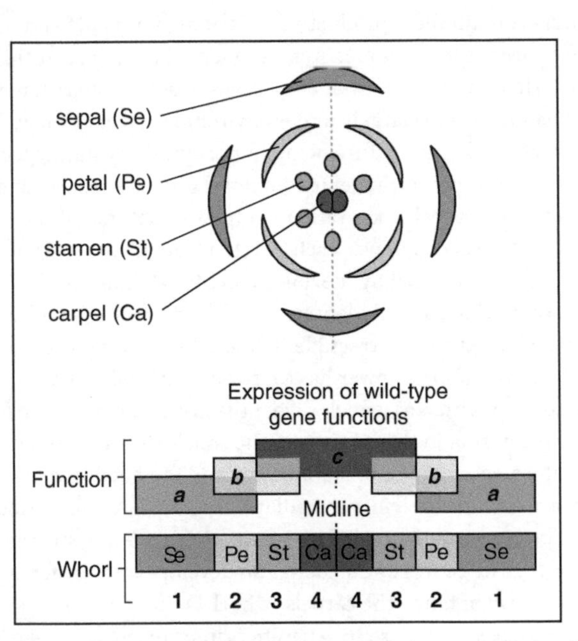

25. Flower development. The genes active in region a specify sepals, regions a and b together specify petals, regions b and c stamens, and those in region c specify carpels

segment identity along the insect's body. In detail, however, there are many differences, and quite different genes are involved. In essence, the floral meristem is divided by the expression patterns of the homeotic genes into three concentric overlapping regions, *a*, *b*, and *c*, which partition the meristem into regions corresponding to the four whorls. Each of the *a*, *b*, and *c* regions corresponds to the zone of action of one class of homeotic genes and the particular combinations of *a*, *b*, and *c* functions give each whorl a unique identity and so specify organ identity. Various studies have shown that different meristem layers communicate with each other during flower development and that transcription factors can move between cells, which does not occur in animal development.

Chapter 8
Nervous system

The nervous system is the most complex of all the organ systems in the animal embryo. In mammals, for example, billions of nerve cells (neurons) develop a highly organized pattern of connections, creating the neuronal network that makes up the functioning brain and the rest of the nervous system. There are also an equivalent number of supporting cells (glia) such as Schwann cells, which insulate nerve cells. As we have seen, during gastrulation, the ectoderm in the dorsal region of the vertebrate embryo becomes specified as the neural plate and the neural plate forms the neural tube, from which the brain develops, while the spinal cord forms more posteriorly. The neural tube throws off neural crest cells, which migrate throughout the body to give rise to neurons and other cell types. The nervous system must develop in the correct relationship with other body structures, such as to the skeleton and muscular system, whose movement it controls.

The induction of neural tissue from ectoderm was first indicated by the Spemann organizer-transplant experiment in frogs. A partial secondary embryo develops when one small region, the Spemann organizer, of an early embryo is grafted onto another embryo at the same stage and a nervous system develops from the host ectoderm that would normally have formed ventral epidermis. An enormous amount of effort was devoted in the 1930s and 1940s to trying to identify the signals involved in neural induction in amphibians. A key discovery was the finding that the BMP (bone morphogenetic protein) inhibitor Noggin, the first secreted protein that was isolated from the Spemann's organizer, could induce neural differentiation in ectoderm explants from frog embryos. The results suggested that neural plate could only develop if BMP signalling is absent. These observations led to the so-called 'default model' for neural induction in the frog. This proposed that the default state of the dorsal ectoderm is to develop as neural tissue, but that this pathway is blocked by the presence of BMP, which promotes it to develop as epidermis. The role of the Spemann organizer is to lift this block by producing proteins that inhibit BMP activity. But the default model was not the complete answer, as neural development in both the frog and the chick also requires other proteins, even when BMP inhibition is lifted by the presence of Noggin. Neural induction is therefore a complex multistep process. An essential similarity in the mechanism of neural induction among vertebrates is likely, however, as Hensen's node from a chick embryo can induce neural gene expression in frog ectoderm, which suggests that the inducing signals have been conserved in evolution.

The nervous system is initially patterned by signals from the underlying mesoderm, and pieces of anterior mesoderm induce a head with a brain, whereas posterior pieces induce a trunk with a spinal cord. Both qualitative and quantitative differences in signalling by the mesoderm can account for antero-posterior neural patterning. Quantitative differences in protein signalling are present along the body axis, with the highest level at the posterior end of the embryo which gives prospective neural tissue a more posterior identity. Hox genes are expressed along the spinal cord and give neurons a positional identity.

There are many hundreds of different types of neurons, differing in identity and the connections they make, even though many may look quite similar (Figure 26). Neurons send out long processes from the cell body and these must be guided

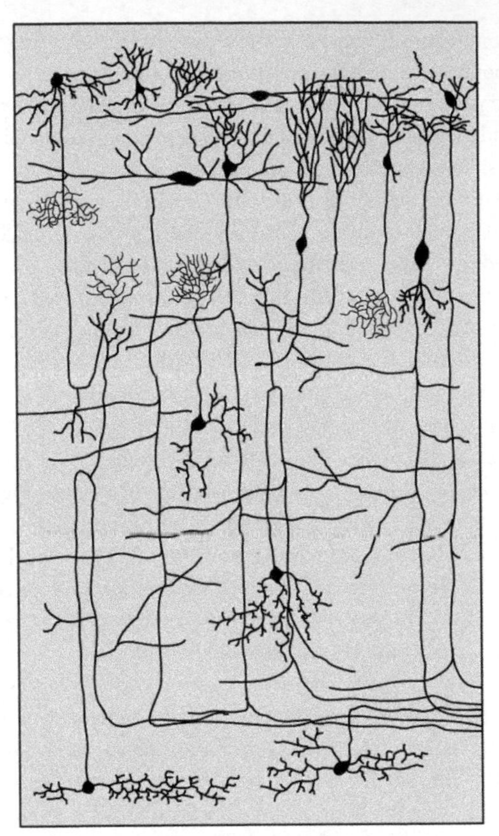

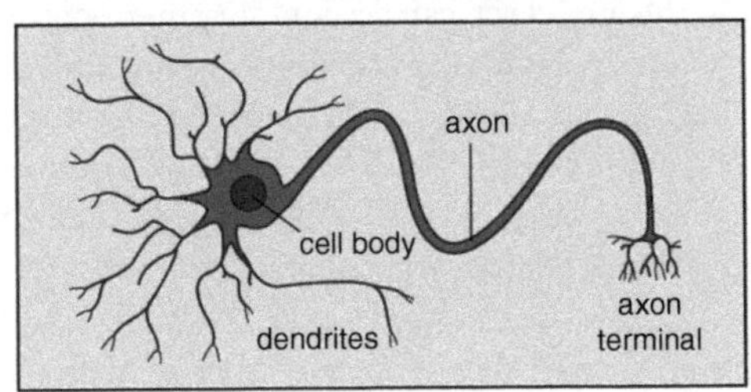

26. Neurons come in many shapes and sizes. In the brain there are many more neural connections than shown here. A single neuron consists of a small rounded cell body, which contains the nucleus, and from which extend a single axon and a 'tree' of much-branched dendrites. Signals from other neurons are received at the dendrites

to find their targets. Neurons send electrical signals (the nerve impulse) down an extension (the axon) which can be very long and which signals to muscles and other neurons. Neurons connect with each other and with other target cells, such as muscle, at specialized junctions known as synapses. A neuron receives input from other neurons through its highly branched short extensions, and if the signals are strong enough to activate the neuron it generates a new electrical signal—a nerve impulse, or action potential—at the cell body. This electrical signal is then conducted along the axon to the axon terminal, or nerve ending, which makes a synapse with another neuron or with the surface of a muscle cell. A single neuron in the central nervous system can receive as many as 100,000 different inputs. At a synapse, the electrical signal is converted into a chemical signal, in the form of a chemical neurotransmitter such as acetylcholine, which is released from the nerve ending and acts on receptors in the membrane of the opposing target cell to generate or suppress a new electrical signal. The nervous system can only function properly if the neurons are correctly connected to one another, and thus a central question surrounding nervous-system development is how the connections between neurons develop with the appropriate specificity. The number of neurons in the human brain seems generally to be estimated at around 100 billion. How many of them have unique or similar identities is not known.

For all its complexity, the nervous system is the product of the same kind of cellular and developmental processes as those involved in the development of other organs. The overall process of nervous-system development can be divided into four major stages: the specification of neural cell identity; the outgrowth of axons to their targets; the formation of synapses with target cells, which can be other neurons, muscle or gland cells; and the refinement of synaptic connections through the elimination of axon branches and cell death (Figure 27).

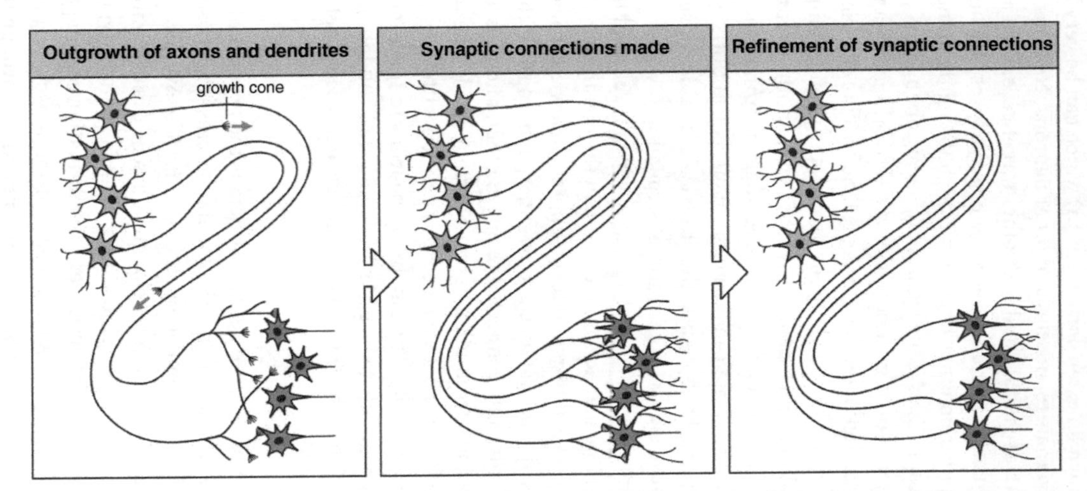

27. Neurons make precise connections with their targets. Axons extend and make numerous contacts that are then refined

Neurons are formed in the proliferative zone of the vertebrate neural tube from multipotent neural stem cells, which give rise to many different types of neurons and to glia. For many years it was thought that no new neurons could be generated in the adult mammalian brain, but the production of new neurons has been demonstrated as a normal occurrence in the adult mammalian brain, and neural stem cells have been identified in adult mammals that can generate neurons.

Future motor neurons are located ventrally, and form the ventral roots of the spinal cord. The neurons of the sensory nervous system develop from neural crest cells. The dorso-ventral organization of the spinal cord is produced by Sonic hedgehog protein signals from ventral regions such as the notochord. Sonic hedgehog forms a gradient of activity from ventral to dorsal in the neural tube, and acts as the ventral patterning positional signal. As well as being organized along the dorso-ventral axis, neurons at different positions along the antero-posterior axis of the spinal cord become specified to serve different functions. The antero-posterior specification of neuronal function in the spinal cord was dramatically illustrated some 40 years ago by experiments in which a section of the spinal cord that would normally innervate wing muscles was transplanted from one chick embryo into the region that normally serves the legs of another embryo. Chicks developing from the grafted embryos spontaneously activated both legs together, as though they were trying to flap their wings, rather than activating each leg alternately as if walking. These studies showed that motor neurons generated at a given antero-posterior level in the spinal cord had intrinsic properties characteristic of that position. The spinal cord becomes demarcated into different regions along the antero-posterior axis by combinations of expressed Hox genes. A typical vertebrate limb contains more than 50 muscle groups with which neurons must connect in a precise pattern. Individual neurons express particular combinations of Hox genes, which determine which muscle they will innervate. So all together, expression of genes resulting from dorso-ventral position together with those resulting from antero-posterior position confers a virtually unique identity on functionally distinct sets of neurons in the spinal cord.

The working of the nervous system depends on the formation of neuronal circuits, in which neurons make numerous and precise connections with each other. A feature of development that is unique to the nervous system is the outgrowth and guidance of axons, long extensions from the nerve cell's body to their final targets. An early event is the extension by the nerve cell of its axon, which is due to the growth cone located at the tip of the axon. The growth cone is specialized for both movement and for sensing its environment for guidance cues. The growth cone can continually extend and retract filopodia at its leading edge, making and breaking connections with the underlying substratum to pull the axon tip forward. The growth cone thus guides axon outgrowth, and is influenced by the contacts the filopodia make with other cells and with the surface over which it moves. In general, the growth cone moves in the direction in which its filopodia make the most stable contacts. In the chick embryo, when the motor neuron axons enter the developing limb bud they are all mixed up in a single bundle. At the base of the limb bud, however, the axons separate out. Even when the axon bundles are made to enter in reverse order, the correct relationship between motor neurons and muscles was achieved. However, many motor neurons make no connections and, as we shall see, they will die.

A complex task for the developing nervous system is to link up the sensory receptors that receive signals from the outside world with their targets in the brain that enable us to make sense of these signals. A characteristic feature of the vertebrate brain is the presence of topographic maps so that neurons from one region of the sensory nervous system project in an ordered manner to a specific region of the brain. The highly organized projection of neurons from the eye via the optic nerve to the brain is one of the best models we have to show how topographic neural projections are made. There are around 126 million individual photoreceptor cells in a human retina, and each of those photoreceptor cells is continuously recording a minute part of the eye's visual field; these signals must be sent to the brain in an orderly manner. Photoreceptor cells indirectly activate individual neurons whose axons are bundled together and exit the eye as the optic nerve; the optic nerve from each human eye, containing over a million neurons, maps in a highly ordered manner onto a specific region of the brain, the tectum (Figure 28). This occurs with a highly ordered correspondence between a position on the retina and one on the tectum. Each retinal neuron carries a chemical label that enables it to connect reliably with an appropriately chemically labelled cell in the tectum. It is thought that graded spatial distributions of a relatively small number of factors on the tectum cells provide positional information, which can be detected by the retinal axons. The spatially graded expression of another set of factors on the retinal axons would provide them with their own positional information. The development of the projection from the eye to the tectum could thus, in principle, result from the interaction between these two gradients. This map is initially rather coarse-grained, in that axons from neighbouring cells in the retina make contacts over a large area of the tectum. Fine-tuning of the map results from the withdrawal of axon terminals from most of the initial contacts, and requires neural activity due to normal vision. If a frog eye is rotated through 180 degrees, the axons find their way back to the tectum and then, for that eye, the frog's world is turned upside down.

Neuronal death is very common in the developing vertebrate nervous system; too many neurons are produced initially and only those that make appropriate connections survive. Some 20,000 motor neurons are formed in the segment of the spinal cord that provides connections to chick leg muscles, but about half of them die soon after they are formed. Survival of a motor neuron depends on its establishing contacts with a muscle cell. Once a contact is established, the neuron can activate the muscle, and this

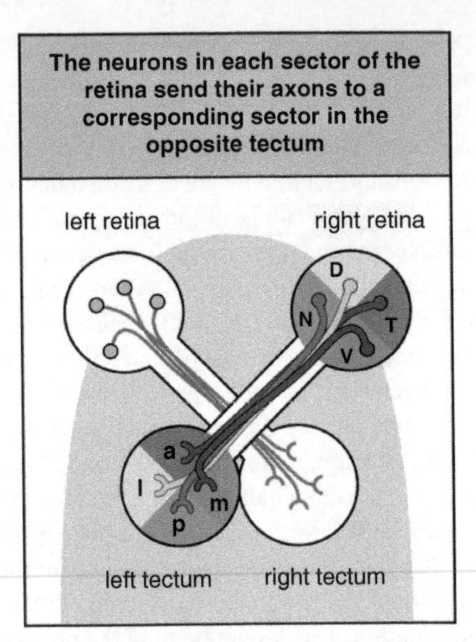

28. Neural connections between the retina and the tectum in the frog. For example, p in the left tectum connects with N in the right retina

is followed by the death of a proportion of the other motor neurons that are approaching the muscle cell and do not make contact. Even after neuromuscular connections have been made, some are subsequently eliminated. At early stages of development, single muscle fibres are contacted by axons from several different motor neurons. With time, most of these connections are eliminated, until each muscle fibre is innervated by the axon terminals from just one motor neuron. This is due to competition between the synapses, with the most powerful input to the target cell destabilizing the less powerful inputs to the same target.

Chapter 9
Growth, cancer, and ageing

Development does not stop once the embryonic phase is complete. Most, but by no means all, of the growth in animals and plants occurs in the post-embryonic period, when the basic form and pattern of the organism has already been established. The basic patterning is done on a small scale, over dimensions of less than a millimetre. In animals most growth continues after birth. In some vertebrates, such as mammals, considerable growth occurs during a late embryonic period, while the embryo is still dependent on maternal resources. Growth is a central aspect of all developing systems, determining the final size and shape of the organism and its parts. Growth of the different parts of the body is not uniform, and different organs grow at different rates. After 9 weeks of embryonic development, the head of a human embryo is more than a third of the length of the whole embryo, whereas at birth it is only about a quarter. After birth, the rest of the body grows much more than the head, which is only about an eighth of the body length in the adult. In mammals, including humans, inadequate nutrition of the embryo not only has direct effects on embryonic and foetal growth but can have serious effects in adult life—there is an increased risk of developing coronary heart disease, stroke, or type 2 diabetes.

Growth can be due to an increase of cell proliferation, cell enlargement without division, or by accretion of extracellular material, such as bone matrix secreted by the cells. Some growth is through a combination of cell proliferation and cell enlargement. For example, the cells in the lens of the eye are produced by cell division, whereas their differentiation involves considerable enlargement. The growth program—that is, how much an organism or an individual organ grows and responds to factors like hormones—may be specified at an early stage in development. As noted earlier, unlike the situation in animals where the embryo is essentially a miniature version of the free-living larva or adult, plant embryos bear little resemblance to the mature plant. Plant growth is achieved by cell division in meristems and organ primordia, followed by irreversible cell enlargement, which achieves most of the increase in size.

Growth hormone is essential for the growth of humans and other mammals after birth. Within the first year of birth, the pituitary gland begins to secrete growth hormone. A child with insufficient growth hormone grows less than normal, but if growth hormone is given regularly, normal growth is restored. In this case, there is a catch-up phenomenon, with a rapid initial response that tends to restore the growth curve to its original trajectory. During the first year after birth growth in height occurs at a rate of about 2 cm per month. The growth rate then declines steadily until the start of a characteristic adolescent growth spurt at puberty, which involves sexual maturation, at about 11 years in girls and 13 years in boys. In pygmies, sexual maturation at puberty is not accompanied by this adolescent growth spurt, hence their characteristic short stature. We still do not know the cellular basis for this.

In mammals, skeletal and heart muscle cells and neurons never divide again once differentiated, although they do increase in size. Neurons grow by the extension and growth of axons and smaller extensions, whereas muscle growth involves an increase in mass, as well as the fusion of satellite cells to pre-existing muscle fibres, providing additional nuclei to support the increase in size. The increase in the length of a muscle fibre depends on the growth of the long bones putting tension on the muscle through its tendons. One can thus see how bone and muscle growth are mechanically coordinated.

Numerous extracellular signalling proteins that can stimulate or inhibit cell proliferation have been discovered. Some cells must receive signals, such as growth factors, not only for them to divide, but also simply to survive. In the absence of all growth factors, such cells commit suicide by apoptosis, as a result of activation of an internal cell death program. There is a significant amount of cell death in all growing tissues, so that overall growth rate depends on the rates of both cell death and cell proliferation.

Organ size in vertebrates can be determined by both internal developmental programs, and by extracellular factors that stimulate or inhibit growth, but the relative importance of these two mechanisms in different organs varies a good deal. The liver, for example, has excellent powers of regeneration in both embryo and adult, whereas the pancreas does not. If a proportion of the liver precursor cells in the embryo are destroyed, the embryonic liver grows back to a normal size, indicating that it does not arise from a fixed number of progenitor cells. The liver secretes some factors that stimulate cell proliferation and others that potentially inhibit growth. When the liver gets to a certain size, the concentration of inhibitory factors in the circulation is sufficient to stop further growth—an example of negative feedback determining the size of an organ. By contrast, if some of the progenitor cells of the pancreas in a mouse embryo are destroyed once the pancreatic 'bud' has formed, a smaller than normal pancreas develops. The size of the embryonic pancreas seems, therefore, to be largely under internal control. Another organ with internal growth control is the thymus gland. If multiple foetal thymus glands are

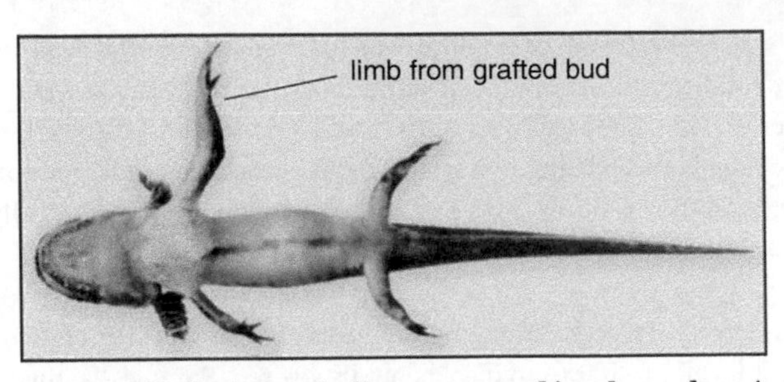

29. The size of limbs is genetically programmed in salamanders. An embryonic limb bud from a large species of salamander grafted to the embryo of a smaller species grows much larger than the host limbs

transplanted into a developing mouse embryo, each one grows to full size. A classic illustration of an internal growth program comes from grafting limb buds between large and small species of salamanders (Figure 29). A limb bud from the larger species grafted to the smaller species initially grows slowly, but eventually ends up at its normal size, which is much larger than any of the limbs of the host.

Each of the cartilaginous elements in the embryonic limb has its own growth program. In the chick embryonic wing, the cartilaginous elements representing the long bones—the humerus and the ulna—are initially similar in size to the elements in the wrist. Yet, with growth, the humerus and ulna increase many times in length compared with the wrist bones before bone formation starts. These growth programs are specified when the elements are initially patterned and involve both cell multiplication and matrix secretion.

An important aspect of post-embryonic vertebrate growth is the growth of the long bones of the limbs like the humerus, radius, and ulna. These long bones are initially laid down as cartilaginous elements and have two internal regions near each end—the growth plates—in which growth occurs; this growth leads to the limb becoming some hundred times longer. In the growth plates, the cartilage cells are arranged in columns, and various zones can be identified. At the site near the end of the bone is a narrow zone that contains stem cells. Next is a proliferative zone of cell division, followed by a zone of maturation, in which the cartilage cells increase in size. Finally, there is a zone in which the cartilage cells die and are replaced by bone. It is the cell divisions and cell enlargement that extends the bone while the growth plate remains the same size. Different bones grow at different rates, and this can reflect the size of the proliferative zone, the rate of proliferation, and the degree of cell enlargement in the growth plate. Growth hormone can stimulate bone growth by acting on the growth plates.

Growth of a bone ceases when the growth plate ossifies, and this occurs at different times for different bones. The timing of growth cessation in the growth plate appears to be intrinsic to the plate itself rather than to hormonal influence. Growth cessation may be due to the cartilage stem cells having only a finite potential for division. In view of the complexity of the growth plate, it is remarkable that our arms on opposite sides of the body can grow for some 15 years independently of each other, and yet eventually match to an accuracy of about 0.2%. How this precision is achieved is not understood.

The fly wing imaginal disc has proved an interesting model system for studying how organ size might be determined. At its formation, the wing disc is initially composed of about 40 cells, and normally grows in the larva to about 50,000 cells. Cell division occurs throughout the disc, and then ceases uniformly when the correct size is reached. The final size of the wing does not depend on the imaginal disc undergoing a fixed number of cell divisions, or attaining a particular number of cells. Instead, final size seems to be controlled by some mechanism that monitors the overall size of the developing wing disc and adjusts cell division and cell size accordingly. Experiments show that there is no restriction on how much of the wing a given cell can make; the descendants of a single cell can contribute from one-tenth to as much as half of the wing. Competition between cells occurs during normal wing growth, and the final size of the wing is achieved through a balance between cell division and apoptosis.

There is evidence that the final size of fly imaginal discs, and thus of the adult organs, might be determined by a molecular gradient across the disc that could be formed as a result of previous patterning. The basic idea is that when the disc is small, the gradient is steep and the steepness of the gradient in some way promotes growth. As the organ grows, the gradient flattens out and growth declines and finally terminates.

Humans are born with a certain number of fat cells, with females generally having more than males. The number of fat cells increases throughout late childhood and early puberty and after this normally remains fairly constant, but an increase in their number can lead to obesity. Although much obesity in children and adults is due to overeating and lack of exercise, early developmental nutritional experience and genetic background can also contribute. Obesity is associated with numerous diseases in later life, including type 2 diabetes and heart disease. Obesity represents both greater numbers of fat cells and excessive deposition of fat in these cells, which increases cell size. Once fat cells develop in the body, they remain there for life and they seldom die. Overweight people with extra fat cells can shrink the size of the cells and lose weight with dieting and exercise, but the fat cells themselves do not disappear, and are only too ready to start accumulating excess fat again.

Animals with a larval stage not only grow in size, but also undergo **metamorphosis**, in which the larva is transformed into the adult form. Metamorphosis often involves a radical change in form and the development of new organs. When an insect larva has reached a particular stage it does not grow and moult any further, but undergoes a radical metamorphosis into the adult form. Metamorphosis occurs in many animal groups. In both insects and amphibians, environmental cues, such as nutrition, temperature, and light, as well as the animal's internal developmental program, control metamorphosis through their effects on hormone-producing cells in the brain. The hormone ecdysone promotes metamorphosis in fly larvae. The expression of at least several hundred genes is altered during fly metamorphosis.

Cancer

Cancer is a major perturbation of normal cellular growth that results from certain mutations in body cells. Creating and maintaining tissue organization requires strict controls on cell division, differentiation, and growth. In cancer, cells escape from these normal controls and proceed along a path of uncontrolled growth and migration that can kill the organism. There is usually a progression from a benign localized growth to malignancy, in which the cells undergo metastasis and migrate to many parts of the body where they continue to grow. Cancers derive from a single abnormal cell that has acquired a number of mutations. The progression of a mutant cell to becoming a tumour-producing cell is an evolutionary process, involving both further mutations and selection of those cells best able to proliferate. The cells most likely to give rise to cancer are those that are undergoing continual division, such as stem cells. Because they replicate their DNA frequently, they are more likely than other cells to accumulate mutations that arise from errors in DNA replication. In almost all cancers, the cancer cells are found to have a mutation in one or more, and usually many, genes. Particular genes in which mutations can contribute to cancer formation have been identified in humans and other mammals. There are also the tumour suppressor genes, in which inactivation or deletion of both copies of the gene is required for a cell to become cancerous.

When an animal cell duplicates, it goes through a fixed sequence of stages called the **cell cycle**. The cell grows in size, the DNA is replicated, the replicated chromosomes are separated by a process known as mitosis, and the cell then divides to give two daughter cells. Once a cell has entered the cell cycle it will continue through and complete the cycle without needing any further external signals. Transitions into successive phases are marked by cell-cycle checkpoints at which the cell monitors progress to ensure, for example, that an appropriate size has been reached, that DNA replication is complete, and that any DNA damage has been repaired. If such criteria are not met, progress into the next stage is delayed until all the necessary processes have been completed. If the cell has suffered some damage that cannot be repaired, the cell cycle will be arrested and the cell will usually undergo apoptosis. The product of tumour suppressor gene *p53* is involved in this checking process.

The tumour suppressor gene *p53* plays a key role in preventing many cancers from developing, and about half of all human tumours contain a mutated form of *p53*. When cells are exposed to agents that damage DNA, then *p53* is activated and arrests the cell cycle, giving the cell time to repair the DNA. The *p53* protein thus prevents the cell from replicating damaged DNA and giving rise to mutant cells. If the damage is too severe to be repaired, *p53* will cause the cell to die by apoptosis. The mutant forms of *p53* found in many cancers do not promote apoptosis, and so the affected cells are more likely to accumulate mutations.

A major feature of cancer is the failure of tumour cells to differentiate properly. The majority of cancers—over 85%—occur in stem cells in cell sheets like the lining of the gut and in lungs where cells are constantly being renewed by division and differentiation of stem cells. Normally, cells generated by stem cells continue to divide for a little time until they undergo differentiation, when they stop dividing. By contrast, cancerous cells continue to divide, although not necessarily more rapidly, and usually fail to differentiate. Another feature of cancer cells, unlike developing cells, is that when they divide, they are genetically unstable which can make them more malignant; the gain or loss of chromosomes is common in solid tumours. The failure of cancer cells to differentiate is also clearly seen in certain cancers of white blood cells. Several types of leukaemia are caused by cells continuing to proliferate instead of differentiating.

Most cancer-related deaths result from tumours that have spread from their site of origin to other tissues, a process known as metastasis. Central to metastasis is the ability of tumour cells to change from being static in a sheet of cells to becoming migrating cells. If the migrating cells enter the blood stream, they can be carried a long way from their origin. Tumours can also attract blood vessels, which enables them to grow more easily.

Ageing

Most organisms are not immortal, even if they escape disease or accidents, because with ageing comes an increasing impairment of physiological functions, which reduces the body's ability to deal with a variety of stresses, and an increased susceptibility to diseases, which can lead to death. Although individuals may vary in the time at which particular aspects of ageing appear, the overall effect is summed up as an increased probability of dying in most animals. But there is little evidence that ageing contributes to mortality in the wild; for example, more than 90% of wild mice die during their first year, several years before ageing affects them. However, elephants can die when old if their tusks are worn out.

Ageing is not part of the organism's developmental program. Rather, ageing is the outcome of an accumulation of damage in cells with time, that eventually outstrips the ability of the body to repair itself, and so leads to the loss of essential functions. It is essentially the result of wear and tear. However, it is fundamental that germ cells do not age as this would prevent reproduction. There is clear evidence that ageing is under genetic control, as different animals age at vastly different rates, as shown by their different lifespans. An elephant, for example, is born after 21 months' embryonic development, and at that point shows few, if any, signs of ageing, whereas a 21-month-old mouse is already well into middle age and just beginning to show signs of senescence. The genetic control of ageing can be understood in terms of the 'disposable soma' theory. This theory puts ageing in the content of evolution, proposing that natural selection tunes the life history of the organism so that sufficient resources are invested in maintaining the cellular repair mechanisms that prevent ageing, at least until the organism has reproduced and cared for its young. Then the organism is disposable, as evolution only cares about reproduction. Cells have numerous mechanisms to delay ageing, which are quite similar to the mechanisms used to prevent malignant transformation. These cellular mechanisms protect the cell from internal damage by reactive chemicals and routinely repair damage to DNA, which is occurring continually in living cells even when they are not actively dividing.

Model animal organisms have been invaluable in investigating what determines ageing and lifespan. These include the nematode worm and the fruit fly, which have short lifespans, and the mouse. Even single-cell organisms like bacteria and yeast age; the parent cell, when it divides, provides a smaller, and essentially younger, offspring cell. Recent landmark molecular genetic studies have identified an evolutionarily conserved biochemical pathway in which an insulin-like growth factor pathway plays a key role and that regulates lifespan in the nematode, fruit fly, and rodents and probably in humans. Reduction of the activity of this pathway appears to increase lifespan and enhance resistance to environmental stress. Genetic variation within the *FOXO3A* gene—the names given to genes can be quite weird—can reduce the pathway's activity and is strongly associated with human longevity.

That there is a limit to the number of times some cells can divide in culture was discovered by Leonard Hayflick in 1965, when he demonstrated that normal human body cells, like fibroblasts, in a cell culture divide about 52 times, but the number is less when the cells are taken from older individuals. There is no such limit for germ cells, cancer cells, or embryonic stem cells. The explanation for the decline in cellular division of body cells in culture with age appears to be linked to the fact that the telomeres, non-coding regions at the ends of chromosomes, get progressively shorter as cells divide. If the telomere gets too short, the cell can no longer divide. The shortening is due to the absence of the enzyme telomerase, which makes the telomere grow back to its normal length after each division. This enzyme is normally expressed only in cells that have to be prevented from ageing, such as germ cells in the testis and ovary, and certain adult stem cells such as those that replace cells in the skin and gut. Cancer cells all have telomerase.

Chapter 10
Regeneration

Regeneration is the ability of the fully developed organism to replace tissues, organs, and appendages. Some amphibians like salamanders show a remarkable capacity for regeneration, being able to regenerate complete new tails and limbs as well as some internal tissues. Some insects and other arthropods can regenerate lost appendages, such as legs. Another striking case of regeneration in vertebrates is the zebrafish, which can regenerate the heart after removal of part of the ventricle. The regenerative powers of mammals are much more restricted. The mammalian liver can regrow if a part of it is removed, and fractured bones mend by a regenerative process. The simple aquatic organism *Hydra* and planarians (flatworms) have major regeneration abilities.

What is the mechanism for regeneration and why are some animals able to regenerate and others not? Understanding regeneration could lead to progress in the development of medical ways of repairing tissues such as the mammalian heart and the spinal cord. A distinction has been drawn between two types of regeneration. In epimorphosis, regeneration involves growth of a new, correctly patterned structure as a limb. In morphallaxis, there is little new cell division and growth, and regeneration of structure occurs mainly by the repatterning of existing tissue; regeneration of the head in *Hydra* is a good example. In epimorphosis, new positional values are linked to growth from the cut surface, while in morphallaxis a new boundary region is first established at the cut, then new positional values are specified in relation to them.

The amputation of a salamander limb is followed by a rapid migration of epidermal cells from the edges of the wound to form a covering over the wound surface. A mass of cells called the blastema then forms under the epidermal cap and this is what gives rise to the regenerated limb. The blastema is formed from cells beneath the wound epidermis that lose their differentiated character and start to divide, eventually forming an elongated cone. As the limb regenerates over a period of weeks, the blastema cells differentiate into cartilage, muscle, and connective tissue. The blastema cells are derived locally from the mesenchymal tissues of the stump, close to the site of amputation.

Do the cells differentiating into cartilage and muscle in the regenerating newt limb remain true to type, or do cells differentiate into another type of cell entirely? Can dedifferentiated skeletal muscle cells in the stump re-differentiate as cartilage, for example? Recent experiments in the regenerating limb showed that the blastema cells did not revert to a pluripotent state, but retained a restricted developmental potential related to their origin. The fate of individual tissue types in a regenerating limb was traced by producing animals that expressed green fluorescent protein in all their cells. A patch of tissue was transplanted from these animals into the forelimbs of an uncoloured animal. The forelimb was then amputated across the site of the transplant, and the fate of the glowing green transplanted cells could then be traced as the limb regenerated. The cells retained their identity. The classic example of cells differentiating into a quite different cell type occurs in lens regeneration in the adult newt eye. When the lens is completely removed by surgery, a new lens regenerates from the pigmented epithelium of the iris.

Growth of the blastema is dependent on its nerve supply. In amphibian limbs in which the nerves have been cut before amputation, a blastema forms but fails to grow. The nerves have no influence on the character or pattern of the regenerated structure but the nerves have been shown to provide an essential growth factor. A striking instance of the influence of nerves on regeneration of amphibian limbs is that if a major peripheral nerve such as the sciatic nerve is cut and the branch inserted into a wound on a limb or on the surface of the adjacent flank, a supernumerary limb develops at that site. An interesting phenomenon, as yet unexplained, is that if embryonic newt limbs are denervated very early in their development and so do not become exposed to the influence of nerves, they can regenerate in the complete absence of any nerve supply.

Regeneration always proceeds in a direction distal to the cut surface. If the hand is amputated at the wrist, only the carpals and digits are regenerated, whereas if the limb is amputated through the middle of the humerus, everything distal to the cut is regenerated. Positional value along the proximo-distal axis is therefore of great importance, and is at least partly retained in the blastema. The blastema has considerable independent developmental potential. If it is transplanted to a neutral location that permits growth, it gives rise to a regenerated structure appropriate to the position from which it was taken.

The ability of cells to recognize a discontinuity in positional values is illustrated by grafting a distal blastema to a proximal stump. In this experiment, the forelimb's stump and blastema have different positional values, corresponding to shoulder and wrist, respectively. The result is a normal limb in which structures between the shoulder and wrist have been generated by growth between the two regions, predominantly from the proximal stump.

A fundamental question in any discussion of pattern formation is the molecular basis of the proposed positional information. A major advance has been the identification of a cell-surface protein Prod1, which is expressed in a graded manner along the proximo-distal axis—shoulder to the hand—of the newt limb. Proximo-distal position of the blastema can be made more proximal by treatment with retinoic acid, which increases the concentration of Prod1. Exposing a regenerating limb amputated at the wrist to retinoic acid results in the positional values of the blastema becoming proximalized because Prod1 is increased; the limb regenerates as if it had originally been amputated at a much more proximal site and almost a whole limb grows from the wrist (Figure 30).

The legs of some insects, such as the cockroach and cricket, can regenerate. Regeneration of insect legs follows an epimorphic process of blastema formation and outgrowth. Intercalation of positional valucs therefore seems to be a general property of epimorphically regenerating systems. When cells with disparate positional values are placed next to one another, intercalary growth occurs in order to regenerate the missing positional values. Intercalation is particularly clearly illustrated by limb regeneration in the cockroach. A cockroach leg is made up of a number of segments, arranged along the proximo-distal axis in

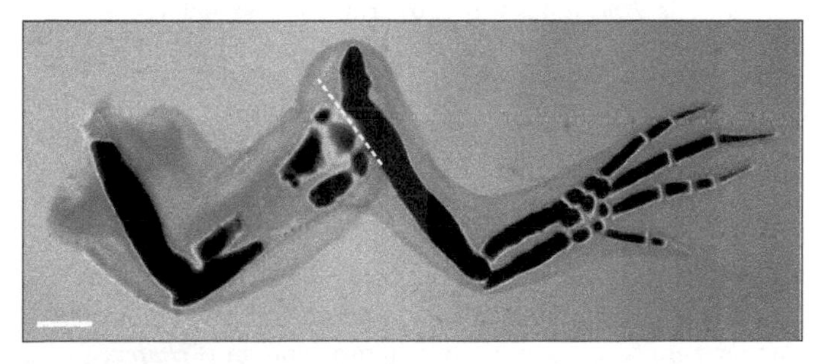

30. The limb was amputated at the hand (dotted line) and treated with retinoic acid during regeneration, which causes the cut surface to have a more proximal positional value and so structures corresponding to a cut at the proximal end of the humerus regenerate

the order: coxa, femur, tibia, and tarsus. Each segment seems to contain a similar set of positional values, and will intercalate missing positional values.

A mammalian system with considerable powers of regeneration in adults is the peripheral nervous system and involves the regrowth of axons but not the replacement of the cells themselves by cell division. Axons of peripheral neurons in adult vertebrates, such as the motor and sensory axons running between the spinal cord and the ends of the limbs, can be up to hundreds of centimetres long. When such an axon is severed, a new growth cone forms at the cut surface and grows down the pathway of the original nerve trunk to make functional connections, leading to an almost complete recovery of function. In the case of motor neurons, the axon terminal finds the site of the original synapse on the muscle cell. By contrast, the central nervous system of adult birds and mammals cannot regenerate.

Hydra is a freshwater simple animal consisting of a hollow tubular body about 0.5 cm long, divided into a head region and a basal region, by which it sticks to a surface. The head consists of a small conical region where the mouth opens, surrounded by a set of tentacles that are used for catching the small animals on which *Hydra* feeds. *Hydra* has only two germ layers: an outer epithelium, which corresponds to the ectoderm; and an inner epithelium lining the gut cavity, which corresponds to the endoderm. The reason for looking at patterning and regeneration in *Hydra* is that it provides insight into an organizer region and developmental gradients that arose early in the evolution of animal development. It is likely that the more complex body patterns of other animals evolved from a simple body plan like that of *Hydra*. Genes identified as important in vertebrate embryo development seem to be involved in regeneration of *Hydra*.

Well-fed *Hydra* are in a dynamic state of continuous growth and pattern formation, and reproduce asexually by budding. But in hard times *Hydra* can reproduce sexually. Budding occurs about two-thirds of the way down the body; the body wall evaginates by a morphogenetic change in cell shape to form a new column which develops a head at the end and then detaches as a small new *Hydra*.

If the body column of a *Hydra* is cut twice transversely to make three pieces, the lower piece will regenerate a head and the upper piece will regenerate a foot. Thus, what structure the cells regenerate at a cut surface depends on their relative position within the regenerating piece. For the middle piece, the cut surface nearest the original head end forms a head—this shows that *Hydra* has a well-defined overall polarity. Regeneration in *Hydra* does not require cell division and new growth and is therefore an example of morphallactic regeneration. When a short fragment of the column regenerates, there is no initial increase in size and the regenerated animal will be a small *Hydra*. Only after feeding will the animal return to a normal size.

At the beginning of the 20th century, it was found that grafting a small fragment of the head region of a *Hydra* into the body region of another *Hydra* induced a new head, complete with tentacles, and a body axis (Figure 31). Similarly, transplantation of a fragment of the basal region induced a new body column with a basal disc at its end. *Hydra* therefore has two organizing regions—one at each end—that give the animal its overall polarity. The head region and the basal disc act as organizing regions like the Spemann organizer in amphibians, and the polarizing regions in vertebrate limb buds. The organizing function of the head region is due to at least two signals it produces that act in graded fashion down the body column: one signal inhibits head formation and the other specifies a gradient in positional values, which determines the level necessary to inhibit head formation.

Provided the level of inhibitor is greater than the threshold set by the positional value, head regeneration is inhibited. Removal of

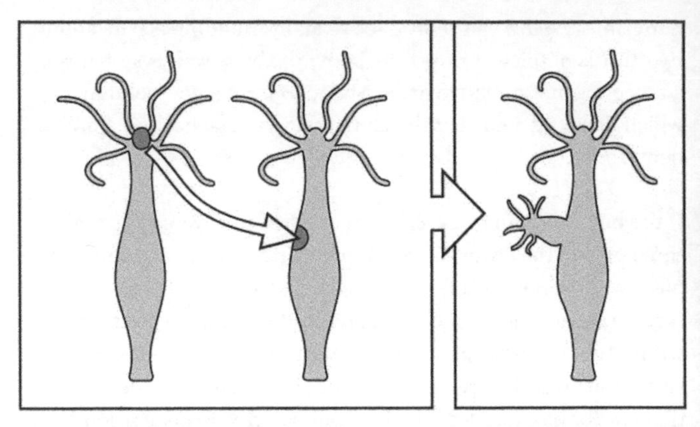

31. A fragment taken from the head region of *Hydra* can induce a new head when grafted into the body of another animal

the head results in a fall in the concentration of inhibitor, and when the inhibitor falls below the threshold concentration set by the local positional value, the positional value increases to that of the head end. Thus, the first key step in this morphallactic regeneration, when the head region is removed, is the specification of a new head region at the cut surface. When the positional value has increased to that of a normal head region, the cells start to make inhibitor and so prevent head formation in other body regions.

Chapter 11
Evolution

Development is a fundamental process in the evolution of multicellular organisms. The evolution of multicellular forms of life, all animals and plants, is the result of changes in embryonic development and these, in turn, are entirely due to changes in genes that control cell behaviour in the embryo and adult. Both changes in the regulation of gene expression in time and space, and mutations in proteins that generate novel protein functions, have played a fundamental role in evolution. It is also true, as the evolutionary biologist Theodosius Dobzhansky once said, that nothing in biology makes sense unless viewed in the light of evolution. Certainly, it would be very difficult to make sense of many aspects of development without an evolutionary perspective. Genetically based changes in development that generated more successful adult forms better adapted to their environment and so able reproduce better, have been selected for during evolution.

Multicellular animals are presumed to descend from a multicellular common ancestor and that had, in turn, evolved from a unicellular organism. As Charles Darwin was the first to realize, evolution is the result of heritable changes in life forms and the selection of those that are best adapted to their environment. 'Darwin's finches' are an excellent example of the evolutionary role of development and of changes in gene expression. Charles Darwin visited the Galapagos Islands in 1835 and collected a group of finches, distinguishing at the time 13 closely related species. What he found particularly striking was the variation in their beaks. The shapes of the beaks reflected differences in the birds' diets and how they got their food. The species with broader, deeper beaks relative to length have now been shown to express higher levels of the bone morphogenetic protein (BMP-4) in the beak growth zone compared with species with long, pointed beaks.

If two groups of animals that differ greatly in their adult structure and habits—such as fishes and mammals—pass through a very similar embryonic stage, this could indicate that they are descended from a common ancestor and, in evolutionary terms, are closely related. All vertebrate embryos pass through a stage at which they all more or less resemble each other (Figure 32). Thus, an embryo's development reflects the evolutionary history of its ancestors. Division of the body into segments, which then diverge from each other in structure and function, is a common feature in the evolution of both vertebrates and arthropods (insects and crustaceans); an example is the development of the somites. In vertebrates, another example of segmented structures are the branchial arches and clefts that are present in all vertebrate embryos, including humans, located just behind the head on either side. These structures are not the relics of the gill arches and gill slits of an adult fish-like ancestor, but represent structures that would have been present in the embryo of the fish-like ancestor of vertebrates as developmental precursors of gill slits and gill arches. During evolution, the branchial arches gave rise both to the gill arches of the primitive jawless fishes and, in a later modification, to gills and jaw elements in later-evolved fishes (Figure 33). With time, the arches became further modified, and in mammals they now give rise to various structures in the face and neck; our jaws come from these arches.

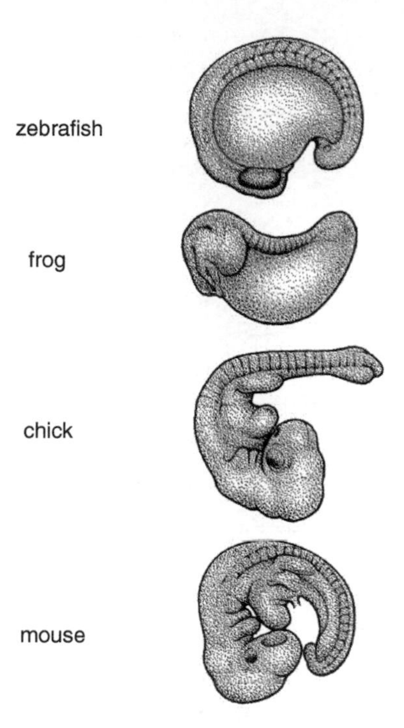

32. Vertebrate embryos at the same stage (tailbud) have similar characters

Evolution rarely, if ever, generates a completely novel structure out of the blue. New anatomical features arise from modification of an existing structure. One can therefore think of much of evolution as a 'tinkering' with existing structures, which gradually fashions something different. It is possible because many structures are modular, that is, animals have anatomically distinct parts that can evolve independently. Vertebrae are modules, for example, and can evolve independently of each other; so too are limbs as we have seen. A nice example of a modification of an existing structure into something quite different is provided by the evolution of the mammalian middle ear, which is made up of three bones—malleus, incus, and stapes—which transmit sound from

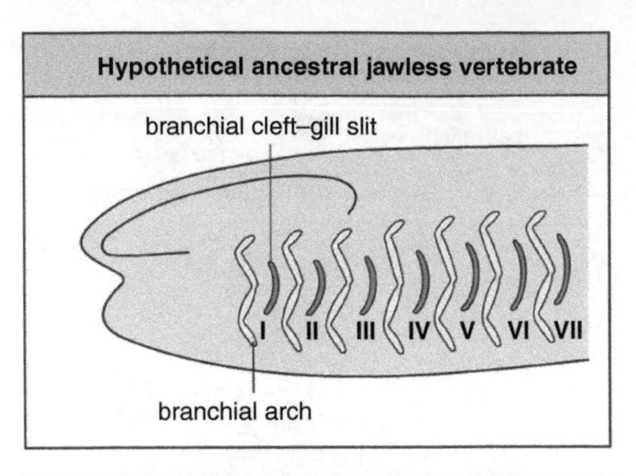

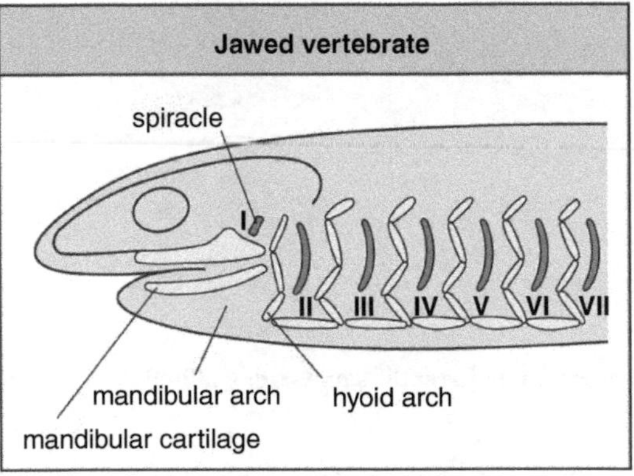

33. Modification of gill arches into jaws during evolution

the eardrum to the inner ear. In the reptilian ancestors of mammals, the joint between the skull and the lower jaw was between the quadrate bone of the skull and the articular bone of the lower jaw, which were also involved in transmitting sound via the stapes. The vertebrate lower jaw was originally composed of several bones but during mammalian evolution one of these bones—the dentary—increased in size and came to comprise the whole lower jaw; the other bones, both quadrate and articular, were no longer attached to it. By changes in their development, the articular and the quadrate in mammals were modified into two bones, the malleus and incus, respectively, whose function was now to transmit sound from the outer ear membrane.

There has been conservation of many developmental mechanisms at the cellular and molecular level among distantly related organisms. For example, widespread use of the Hox gene complexes and the same few families of protein signalling molecules provide excellent examples of this. It is these basic similarities in molecular mechanisms that have made the study of developmental biology so exciting in recent years; it has meant that discoveries of genes in one animal have had important implications for understanding development in other animals. It seems that when a useful developmental mechanism evolved, it was retained and redeployed in very different organisms, and at different times and places in the same organism. Signalling molecules are already present in simple multicellular animals such as the *Hydra*, which arose early in animal evolution.

The largest group of animals is the Bilateria, which include the vertebrates and the arthropods such as the insects and crustaceans. They all have bilateral symmetry across the main body axis in at least some stage of development, and have a characteristic pattern of Hox gene expression. The origin of the ancestor of animals is a tough problem but a possible mechanism will be proposed. The last common ancestor of the bilaterians must already have been quite a complex creature, possessing most of the developmental gene pathways used by existing animals. It is speculated that the ancestor lived some 600 million years ago and would have had flagellate sperm, development through a process of gastrulation, multiple germ layers, neuromuscular and sensory systems, and fixed body axes. There is now one very simple and primitive free-living marine animal, *Trichoplax*, which may be close to the origin. It is composed of just two layers of cells, which form a flat disc with no gut, and it has only four different cell types. It mainly reproduces by fission. Nevertheless, in line with the genomes of other animals, the *Trichoplax* has an estimated 11,500 protein-coding genes, which encode a rich array of transcription factors and signalling proteins, some of which are similar to those in vertebrates.

A major general mechanism of evolutionary change has been gene duplication and divergence. The duplication of a gene, which can occur by a variety of mechanisms during DNA replication, provides the embryo with an additional copy of the gene. This extra copy can diverge in both its coding sequence and its regulatory regions, thus changing its pattern of expression and its protein's downstream targets without depriving the organism of the function of the original gene. The process of gene duplication has been fundamental in the evolution of new proteins and new patterns of gene expression; it is clear, for example, that the different hemoglobins in humans have arisen as a result of gene duplication. The Hox gene complexes provide one of the clearest examples of the importance of gene duplication in developmental evolution. The Hox genes have also evolved by duplication of a single ancestral gene. The simplest Hox gene complexes are found in invertebrates, and comprise a small number of sequence-related genes carried on one chromosome. Vertebrates typically have four sets of Hox genes, carried on four different chromosomes, suggesting two rounds of wholesale duplication of an ancestral Hox gene complex, in line with the generally accepted idea that large-scale duplications of the genome have occurred during vertebrate evolution. The advantage of duplication was that the embryo had more Hox genes to control downstream targets and so could make a more complicated body. The number of vertebrae in a particular region varies considerably among the different vertebrate classes: mammals, with rare exceptions, have 7 cervical vertebrae, whereas birds can have between 13 and 15. How does this difference arise? A comparison between the mouse and the chick shows that the domains of Hox gene expression have shifted in parallel with the change in number of vertebrae in the different regions. Snakes have hundreds of similar vertebrae in their backbones. Hox genes expressed in the thoracic region of four-limbed vertebrates are expressed along much of the body in the python embryo. The expansion of these Hox expression domains is thought to underlie the expansion of rib-bearing vertebrae and the loss of forelimbs in snake evolution.

Good illustrations of the evolution of Hox genes in regional specification are provided by arthropod appendages. Insect fossils display a variety of patterns in the position and number of their paired appendages—principally the legs and wings. Some insect fossils have legs on every segment, whereas others only have legs in a distinct thoracic region. This suggests that the potential for appendage development is present in every segment, even in flies, and is actively repressed in the fly abdomen by Hox genes. It thus seems likely that the ancestral arthropod from which insects evolved had appendages on all its segments. The Hox genes can also determine the nature of an appendage and we have seen how mutations can convert legs into antenna-like structures and an antenna into a leg.

Amphibians, reptiles, birds, and mammals have limbs, whereas fish have fins. The limbs of the first land vertebrates evolved from the pectoral fins of their fish-like ancestors but how fins evolved is far less clear. Nevertheless, the development of these appendages made use of signalling molecules like Sonic hedgehog and transcription factors such as the Hox proteins, which were already being used to pattern the body. The fossil record suggests that the transition from fins to limbs occurred in the Devonian period, between 400 and 360 million years ago, when the fish ancestors living in shallow waters moved onto the land. The proximal skeletal elements of the ancestral fin are probably related to the humerus, radius, and ulna of the limb, and a recent analysis of a fossil *Panderichthys* has shown that the distal region of its pectoral fin contains separate skeletal elements, and so fingers may not be an evolutionary novelty.

To gain insights into the transition from fin to limb, researchers have turned to a modern fish, the zebrafish, in which fin development can be followed in detail and the genes involved can be identified. The fin buds of the zebrafish embryo are initially similar to vertebrate limb buds, but important differences soon arise during development. As in the vertebrate limb bud, the key gene *Sonic hedgehog* is expressed at the posterior margin of the zebrafish fins and the expression pattern of *Hoxd* and *Hoxa* genes is similar to that in vertebrates. The essential difference between fin and limb development is in the distal skeletal elements; in the zebrafish fin bud, a fin fold develops at the distal end of the bud and fine bony fin rays, not digits, are formed within it.

The great range of anatomical specializations that have evolved in the limbs of mammals are due to changes both in limb patterning and in the differential growth of parts of the limbs during embryonic development, but the basic underlying pattern of skeletal elements is maintained. This is an excellent example of the modularity of the skeletal elements. If one compares the forelimb of a bat and a horse, one can see that although both retain the basic pattern of limb bones, each has been modified to provide a specialized function. In the bat, the limb is adapted for flying and the digits are greatly lengthened to support a membranous wing. Because individual structures such as bones can grow at different rates, the overall shape of an organism can be changed substantially during evolution by heritable changes in the duration of growth, which also leads to an increase in overall size of the organism. In the horse, for example, the central toe of the ancestral horse grew faster than the toes on either side, so that it ended up longer than the lateral digits. As horses continued to increase in overall size during evolution, this discrepancy in growth rates resulted in the relatively smaller lateral toes no longer touching the ground because of the much greater length of the central digit. At a later stage in evolution, the now-redundant lateral toes became reduced even further in size.

Many animals have evolved larval forms that have an advantage when it comes to dispersal and feeding, and they then undergo a dramatic change in form—metamorphosis—to reach the adult state. The essence of development is gradual change, yet at metamorphosis there is no gradual continuity between larva and adult. Metamorphosis makes more evolutionary sense, however, if it is assumed that all larval forms evolved by the insertion of the larval stage into the pre-existing development program of a directly developing animal. In many invertebrates, the larva initially resembles the late gastrula stage, which could have given rise to the free-swimming larval form. Metamorphosis brings the larva back into the original developmental program.

Evolution can also adapt the same proteins for quite different purposes. The eye lenses of octopuses and squids and of vertebrates consist of cells packed with crystallin proteins, which give the lens its transparency. The crystallins were originally thought to be unique to the lens and to have evolved for this special function, but more recent research indicates that they are co-opted proteins that are not structurally specialized for lens function, and in other contexts act as enzymes. These examples provide evidence of a key relationship between evolution and development; the gradual change of a structure into a different form. In many cases, however, we do not understand how intermediate forms were adaptive and gave a selective advantage to the animal. Consider, for example, the intermediate forms in the transition of the first branchial arch to jaws; what was the adaptive advantage? The wings of insects evolved from structures used for getting oxygen from the water, so what was their initial advantage when insects left the water? We do not know, and because of the passage of time and our current ignorance of the ecology of ancient organisms we may never know.

If recognizable multicellular animals had evolved by some 600 million years ago, the question still remains of how they evolved from a unicellular ancestor. What had to be invented for the transition from single cells to multicellularity? How did embryonic development from an egg evolve? The key requirements for embryonic development, as we have seen, are a program of gene activity, cell differentiation, cell motility and cohesion. Judging by modern unicellular organisms with a nucleus and mitochondria, the single-celled organism ancestral to animals would have possessed all these features in primitive form, and little new would have had to be invented. One possibility, and it is highly speculative, is that mutations resulted in the progeny of a single-celled organism not separating after cell division, leading to a loose colony of identical cells that occasionally fragmented to give new 'individuals'. One advantage of a colony might originally have been that when food was in short supply, the cells could feed off each other, and so the colony survived. This could have been the origin of multicellularity, and the egg may subsequently have evolved as the cell fed by other cells; in modern sponges, for example, the egg eats neighbouring cells. Once multicellularity evolved, it opened up all sorts of new possibilities, such as cell specialization for different functions. There was also the advantage that all the cells in an embryo had the same genes and this made cooperation and signalling possible.

How gastrulation evolved is also unknown, but it is not implausible to consider a scenario in which a hollow sphere of cells, the common ancestor of all multicellular animals, changed its form to assist feeding. This ancestor may, for example, have sat on the ocean floor, ingesting food particles by phagocytosis. A small invagination developing in the body wall could have promoted feeding by forming a primitive gut. Movement of cilia could have swept food particles more efficiently into this region, where they could be taken up by the cells. Once the invagination formed, it is not too difficult to imagine how it could eventually extend right across the sphere, fuse with the other side, and form a continuous gut, which would be the endoderm. At a later stage in evolution, cells migrating inside between the gut and the outer epithelium would give rise to the mesoderm. Gastrulation provides a good example of developmental changes during evolution. While there is considerable similarity in the process of gastrulation in many different animals, there are also significant differences. But how these evolved and what could have been the adaptive nature of the intermediate forms remains unknown.

Finally, we can contemplate the evolution of our understanding of developmental biology. Progress has been impressive but due to the complexity of cells with all their proteins and other molecules interacting, there is still much to be learned. It is likely that in the next 50 years, given the genes and structure of a fertilized egg, it will be possible to reliably compute the details of that organism's development and just what the adult would be.

Glossary

apoptosis or programmed cell death, a type of cell death that occurs widely during development. In programmed cell death, a cell is induced to commit suicide

cell cycle, the sequence of events by which a cell duplicates itself and divides in two

cleavage, a series of rapid cell divisions without growth that divides the embryo up into a number of small cells following fertilization

epiblast, a group of cells in mouse and chick embryos that gives rise to the embryo proper

gastrulation, the process in animal embryos in which prospective endodermal and mesodermal cells move from the outer surface of the embryo to the inside, where they give rise to internal organs

gene, a region in the DNA of chromosomes that codes for a protein

Hox genes, encode transcription factors involved in patterning

imprinting, the process by which different genes are inactivated during formation of the germ cells (egg and sperm)

induction, the process whereby one group of cells signals to another group of cells and so affects how they will develop

meristems, groups of undifferentiated, dividing cells that persist at the growing tips of plants. They give rise to all the adult structures—shoots, leaves, flowers, and roots

metamorphosis, the process by which a larva is transformed into an adult. It often involves a radical change in form, and the development of new organs, such as wings in butterflies and limbs in frogs

morphogen, any substance active in pattern formation whose spatial concentration varies and to which cells respond differently at different threshold concentrations

morphogenesis, the processes involved in bringing about changes in form in the developing embryo

neurulation, the process in vertebrates in which the ectoderm of the future brain and spinal cord—the neural plate—develops folds that come together to form the neural tube

pattern formation, the process by which cells in a developing embryo acquire identities that lead to a well-ordered spatial pattern

pluripotency, stem cells, such as embryonic stem cells, that can give rise to all types of cells in the body

positional information, the positional value that cells acquire during pattern formation. The cells then interpret this positional value according to their genetic constitution and developmental history, and develop accordingly

regulation, the ability of the embryo to develop normally even when parts are removed or rearranged

stem cell, a cell that retains the capacity to develop into more than one differentiated cell type. Stem cells divide many times and one of the daughter cells remains a stem cell while the other gives rise to a differentiated cell type

totipotency, the capacity of a cell to develop into a new organism

transcription factors, regulatory proteins required to initiate or regulate the transcription of a gene into RNA. Transcription factors act within the nucleus of a cell by binding to specific regulatory regions in the DNA

Further reading

Slack, J.M. 2006 *Essential Developmental Biology* 2nd edn. Wiley–Blackwell

Wolpert, L. and Tickle, C. 2010 *Principles of Development* 4th edn. Oxford University Press

Index

E

F

G

H

I

L

M

N

O

P

Expand your collection of
VERY SHORT INTRODUCTIONS

1. Classics
2. Music
3. Buddhism
4. Literary Theory
5. Hinduism
6. Psychology
7. Islam
8. Politics
9. Theology
10. Archaeology
11. Judaism
12. Sociology
13. The Koran
14. The Bible
15. Social and Cultural Anthropology
16. History
17. Roman Britain
18. The Anglo-Saxon Age
19. Medieval Britain
20. The Tudors
21. Stuart Britain
22. Eighteenth-Century Britain
23. Nineteenth-Century Britain
24. Twentieth-Century Britain
25. Heidegger
26. Ancient Philosophy
27. Socrates
28. Marx
29. Logic
30. Descartes
31. Machiavelli
32. Aristotle
33. Hume
34. Nietzsche
35. Darwin
36. The European Union
37. Gandhi
38. Augustine
39. Intelligence
40. Jung
41. Buddha
42. Paul
43. Continental Philosophy
44. Galileo
45. Freud
46. Wittgenstein
47. Indian Philosophy
48. Rousseau
49. Hegel
50. Kant
51. Cosmology
52. Drugs
53. Russian Literature
54. The French Revolution
55. Philosophy
56. Barthes
57. Animal Rights
58. Kierkegaard
59. Russell
60. Shakespeare
61. Clausewitz
62. Schopenhauer
63. The Russian Revolution
64. Hobbes
65. World Music

66. Mathematics
67. Philosophy of Science
68. Cryptography
69. Quantum Theory
70. Spinoza
71. Choice Theory
72. Architecture
73. Poststructuralism
74. Postmodernism
75. Democracy
76. Empire
77. Fascism
78. Terrorism
79. Plato
80. Ethics
81. Emotion
82. Northern Ireland
83. Art Theory
84. Locke
85. Modern Ireland
86. Globalization
87. The Cold War
88. The History of Astronomy
89. Schizophrenia
90. The Earth
91. Engels
92. British Politics
93. Linguistics
94. The Celts
95. Ideology
96. Prehistory
97. Political Philosophy
98. Postcolonialism
99. Atheism
100. Evolution
101. Molecules
102. Art History
103. Presocratic Philosophy
104. The Elements
105. Dada and Surrealism
106. Egyptian Myth
107. Christian Art
108. Capitalism
109. Particle Physics
110. Free Will
111. Myth
112. Ancient Egypt
113. Hieroglyphs
114. Medical Ethics
115. Kafka
116. Anarchism
117. Ancient Warfare
118. Global Warming
119. Christianity
120. Modern Art
121. Consciousness
122. Foucault
123. The Spanish Civil War
124. The Marquis de Sade
125. Habermas
126. Socialism
127. Dreaming
128. Dinosaurs
129. Renaissance Art
130. Buddhist Ethics
131. Tragedy
132. Sikhism
133. The History of Time
134. Nationalism
135. The World Trade Organization
136. Design
137. The Vikings
138. Fossils

210. Fashion
211. Forensic Science
212. Puritanism
213. The Reformation
214. Thomas Aquinas
215. Deserts
216. The Norman Conquest
217. Biblical Archaeology
218. The Reagan Revolution
219. The Book of Mormon
220. Islamic History
221. Privacy
222. Neoliberalism
223. Progressivism
224. Epidemiology
225. Information
226. The Laws of Thermodynamics
227. Innovation
228. Witchcraft
229. The New Testament
230. French Literature
231. Film Music
232. Druids
233. German Philosophy
234. Advertising
235. Forensic Psychology
236. Modernism
237. Leadership
238. Christian Ethics
239. Tocqueville
240. Landscapes and Geomorphology
241. Spanish Literature
242. Diplomacy
243. North American Indians
244. The U.S. Congress
245. Romanticism
246. Utopianism
247. The Blues
248. Keynes
249. English Literature
250. Agnosticism
251. Aristocracy
252. Martin Luther
253. Michael Faraday
254. Planets
255. Pentecostalism
256. Humanism
257. Folk Music
258. Late Antiquity
259. Genius
260. Numbers
261. Muhammad
262. Beauty
263. Critical Theory
264. Organizations
265. Early Music
266. The Scientific Revolution
267. Cancer
268. Nuclear Power
269. Paganism
270. Risk
271. Science Fiction
272. Herodotus
273. Conscience
274. American Immigration
275. Jesus
276. Viruses
277. Protestantism
278. Derrida
279. Madness
280. Developmental Biology

HIV/AIDS

A Very Short Introduction

Alan Whiteside

HIV/AIDS is without doubt the worst epidemic to hit humankind since the Black Death. The first case was identified in 1981; by 2004 it was estimated that about 40 million people were living with the disease, and about 20 million had died. The news is not all bleak though. There have been unprecedented breakthroughs in understanding diseases and developing drugs. Because the disease is so closely linked to sexual activity and drug use, the need to understand and change behaviour has caused us to reassess what it means to be human and how we should operate in the globalising world. This *Very Short Introduction* provides an introduction to the disease, tackling the science, the international and local politics, the fascinating demographics, and the devastating consequences of the disease, and explores how we have — and must — respond.

> 'It won't make you an expert. But you'll know what you're talking about and you'll have a better idea of all the work we still have to do to wrestle this monster to the ground.'
>
> **Aids-free world website.**

www.oup.com/vsi

EVOLUTION

A Very Short Introduction

Brian Charlesworth and Deborah Charlesworth

In less than 500 years the relentless application of the scientific method of inference from experiment and observation, without reference to religious or governmental authority has completely transformed our view of our origins and relation to the universe.

This book is about the crucial role of evolutionary biology in transforming our view of human origins and relation to the universe, and the impact of this idea on traditional philosophy and religion. The purpose of this book is to introduce the general reader to some of the most important basic findings, concepts, and procedures of evolutionary biology, as it has developed since the first publications of Darwin and Wallace on the subject, over 140 years ago. Evolution provides a unifying set of principals for the whole of biology; it also illuminates the relation of human beings to the universe and each other.

www.oup.com/vsi

ONLINE CATALOGUE

A Very Short Introduction

Our online catalogue is designed to make it easy to find your ideal Very Short Introduction. View the entire collection by subject area, watch author videos, read sample chapters, and download reading guides.